Redefining Standard Model Cosmology

Edited by Brian Albert Robson

Published in London, United Kingdom

IntechOpen

Supporting open minds since 2005

Redefining Standard Model Cosmology
http://dx.doi.org/10.5772/intechopen.75275
Edited by Brian Albert Robson

Contributors
Dietmar E. Rothe, Oscar Castillo Felisola, José Perdiguero, Oscar Orellana, Hector Hortua, Leonardo Castaneda, Minghui Shao, Zhifu Gao, Ho-Ming Mok, Brian Albert Robson

First published in London, United Kingdom, 2019 by IntechOpen
IntechOpen is the global imprint of INTECHOPEN LIMITED, registered in England and Wales, registration number: 11086078, The Shard, 25th floor, 32 London Bridge Street
London, SE19SG – United Kingdom
Printed in Croatia

British Library Cataloguing-in-Publication Data
A catalogue record for this book is available from the British Library

Additional hard and PDF copies can be obtained from orders@intechopen.com

Redefining Standard Model Cosmology
Edited by Brian Albert Robson
p. cm.
Print ISBN 978-1-83880-863-1
Online ISBN 978-1-83880-864-8
eBook (PDF) ISBN 978-1-83880-865-5

For EU product safety concerns:
IN TECH d.o.o., Prolaz Marije Krucifikse Kozulić 3, 51000 Rijeka, Croatia, info@intechopen.com or visit our website at intechopen.com.

Meet the editor

Professor Brian Albert Robson obtained his MSc, PhD, and DSc degrees in Physics from the University of Melbourne, Australia. He is a Fellow of both the Australian Institute of Physics and the UK Institute of Physics. Currently, he is an honorary professor in the Research School of Physics and Engineering, the Australian National University, Canberra. During his academic career, he served for four years as the officer-in-charge of the Australian National University's first computer, for nine years as the head of the Department of Theoretical Physics, and for two years as an associate director of the Research School of Physics and Engineering. Professor Robson has published more than 150 scientific publications mainly in the areas of nuclear physics, particle physics, gravitation, and cosmology.

Contents

Preface

The current standard model of cosmology, also called the concordance cosmological model or the Lambda-CDM model, assumes that the universe was created in the "Big Bang" from pure energy and is now composed of about 5% ordinary matter, 27% dark matter, and 68% dark energy. This cosmological model is based primarily on two theoretical models: (1) the standard model of particle physics, which describes the physics of the very small in terms of quantum mechanics, and (2) the general theory of relativity, which describes the physics of the very large in terms of classical physics. Both these theoretical models are considered to be incomplete in the sense that they do not provide any understanding of several empirical observations, such as the Big Bang, dark matter, dark energy, gravity, and matter-antimatter asymmetry in the universe. In addition, the current standard model of cosmology makes several assumptions involving the homogeneous and isotropic nature of the universe and two unknown entities called dark matter and dark energy. The main aim of this book is to discuss these serious problems that threaten to undermine the current standard model of cosmology.

In the introductory Chapter 1, numerous dubious assumptions of the current standard model are discussed in some detail. The following Chapters 2–6 are divided into two main sections. Section 2 is devoted primarily to alternatives to the Big Bang scenario of the standard model based on modifications of the steady-state models that were popular prior to the discovery of cosmic microwave background radiation. Section 3 contains chapters that discuss modifications to the general theory of relativity. These include suggestions of alternative models that provide a generalization of the field equations of the general theory of relativity and a generalization of the equivalence principle to reconcile quantum mechanics and general relativity.

Brian Albert Robson
Honorary Professor,
Research School of Physics and Engineering,
The Australian National University,
Canberra, Australia

Section 1

Introduction

Chapter 1

Introductory Chapter: Standard Model of Cosmology

Brian Albert Robson

1. Introduction

The current Standard Model of Cosmology (SMC), also called the "Concordance Cosmological Model" or the "ΛCDM Model," assumes that the universe was created in the "Big Bang" from pure energy, and is now composed of about 5% ordinary matter, 27% dark matter, and 68% dark energy [1].

While the SMC is based primarily upon two theoretical models: (1) the Standard Model of Particle Physics (SMPP) [2], which describes the physics of the very small in terms of quantum mechanics and (2) the General Theory of Relativity (GTR) [3], which describes the physics of the very large in terms of classical mechanics; it also depends upon several additional assumptions.

The main additional assumptions of the SMC are: (1) the universe was created in the Big Bang from pure energy; (2) the mass energy content of the universe is given by 5% ordinary matter, 27% dark matter, and 68% dark energy; (3) the gravitational interactions between the above three components of the mass energy content of the universe are described by the GTR; and (4) the universe is homogeneous and isotropic on sufficiently large (cosmic) scales.

Unfortunately, both the SMPP and the GTR are considered to be *incomplete* in the sense that they do not provide any understanding of several empirical observations. The SMPP does not provide any understanding of the existence of three families or generations of leptons and quarks, the mass hierarchy of these elementary particles, the nature of gravity, the nature of dark matter, etc. [4]. The GTR does not provide any understanding of the Big Bang cosmology, inflation, the matter-antimatter asymmetry in the universe, the nature of dark energy, etc.

Furthermore, the latest version of the SMC, the ΛCDM Model is essentially a parameterization of the Big Bang cosmological model in which the GTR contains a cosmological constant, Λ, which is associated with dark energy, and the universe contains sufficiently massive dark matter particles, i.e., "cold dark matter." However, both dark energy and dark matter are simply names describing unknown entities.

The main aim of this Cosmology Book is to discuss the above serious problems that threaten to undermine the foundations of the current SMC.

2. Dubious assumptions of SMC

The current SMC has numerous dubious assumptions that will be discussed in the following. It will be indicated that many of the problems associated with the SMC arise from the dubious assumption that the GTR is valid for all distances within the expanding universe and not just for the very small distances of the Solar System.

In 1916, Einstein published his GTR, which today is still regarded as the best theory of gravity. The GTR, representing gravitational interactions in terms of the geometry of space-time [3], is equivalent to Newtonian gravity provided that the concentration of mass energy is not too great. However, the GTR is clearly superior to Newtonian gravity since it is consistent with special relativity and in addition provides an understanding of several observations unexplained by the Newtonian theory, e.g., the anomalous perihelion advance of the planet Mercury and the deflection of starlight by the Sun during a total eclipse is twice that predicted by the Newtonian theory.

The GTR describes space-time by a metric that determines the distances separating nearby points (stars, galaxies, etc.). The assumption that the metric should be homogeneous and isotropic on large scales uniquely requires that the metric be the Friedmann-Lemaître-Robertson-Walker metric (FLRW metric). Historically, Friedmann in 1922 [5] simplified the field equations of GRT by assuming that the universe is homogeneous and isotropic. In 1927, Lemaître [6] obtained similar results to Friedmann. The majority of the solutions of the Friedmann field equations predict an expanding or a contracting universe, depending upon initial parameters such as the mass-energy of the universe. Both Robertson [7] and Walker [8] considered the theory further and proved that the FLRW metric is the only one that is spatially homogeneous and isotropic.

The use of the FLRW metric assumes that the universe is spatially both homogeneous and isotropic on each reasonably large scale, e.g., on galactic scales. This is a dubious assumption, since recent astronomical observations have shown that the distribution of galaxies is definitely not smooth, displaying filamentary structures separated by regions containing very few galaxies.

In 1929, Hubble [9] discovered that light from remote galaxies was redshifted, implying that these galaxies were receding from the Earth. Hubble observed that there is a linear relationship between the radial speed with which a galaxy recedes from the Earth and its distance from the Earth. The constant of proportionality is known as the Hubble constant. Recently, the International Astronomical Union resolved that from now on, the expansion of the universe be referred to as the Hubble-Lemaître law. If the universe is expanding, this implies that (i) only the expanding solutions of the Friedmann equations are allowed as solutions for the universe and (ii) the universe must have had a very dense and hot beginning. It should be noted that the expanding solutions of Friedmann consider that it is space itself that is expanding and that the galaxies are at rest within the expanding space. Thus, the redshift for each galaxy is a consequence of the wavelength of the light being stretched by the expansion of space and is *not* a normal Doppler redshift.

In 1927, Lemaître noted that an expanding universe could be extrapolated back in time to an originating singular point that has become associated with the notion of the Big Bang. Lemaître called this original very small and compact hot dense universe the "primordial atom" and considered that the present universe arose as a result of the observed expansion.

The prevailing model of the Big Bang is based upon the GTR. According to this theory, extrapolation of the expansion of the universe backward in time yields an infinite mass-energy density and temperature at a finite time, approximately 13.8 billion years ago. Thus the "birth" of the universe appears to be associated with a *singularity*, which describes not only a breakdown of the GTR, but also all the laws of physics. This suggests a dubious assumption associated with the Big Bang hypothesis, indicating that the GTR with the FLRW metric is *not valid* for extremely small regions of space.

On the other hand, the Big Bang scenario has had some success. In 1948, Gamow [10] suggested that the present features of the universe could be understood as a result of the evolutionary development of the universe via expansion from the Big Bang phase. In particular, he suggested that the elements could have been made during the early hot matter-energy phase associated with the Big Bang. It has since been shown that as the initial hot dense mass-energy phase of the universe cooled during the expansion that only several light elements were formed, including hydrogen ($\approx$75%), helium ($\approx$25%), and small amounts of deuterium, lithium, etc.

As the hot dense phase continued to cool down during the expansion, the atomic nuclei of hydrogen, helium, etc. captured electrons, thereby, generating *neutral* atoms. This is estimated to have occurred about 400,000 years after the Big Bang, when photons ceased interacting significantly with matter, leading to the occurrence of the so-called *Cosmic Microwave Background* (CMB) radiation. In 1948, Alpher and Herman [11] calculated the present temperature of this CMB to be about 5 K, remarkably close to the modern value of about 2.73 K, determined by the COBE satellite [12]. In addition, the COBE results showed an extremely isotropic and homogeneous CMB. This led to the need for an *inflationary* phase [13] of strongly accelerated expansion prior to the decoupling of photons from ordinary matter.

In the 1960s, the interpretation of the CMB as the remnant from an early stage of the universe following the Big Bang was challenged by some proponents [14] of the Steady-State Model [15, 16] of the universe. They argued that the microwave background was the result of scattered starlight from distant galaxies. However, the discovery of the CMB radiation in 1964 by Penzias and Wilson [17], especially with the later results from the COBE satellite, which indicated that the CMB spectrum was a thermal black body spectrum, strongly supported the fact that the CMB is a remnant of the Big Bang.

The SMC has two additional major dubious assumptions: the existence of both *dark matter* and *dark energy*, which the SMC claims constitute about 27% and 68%, respectively, of the mass-energy content of the universe.

The notion of "dark matter" arose from observations of large astronomical objects such as galaxies and clusters of galaxies, which displayed gravitational effects that could not be accounted for by the visible matter: stars, hydrogen gas, etc., assuming the validity of Newton's universal law of gravitation. In particular, the observations of Rubin et al. [18], who measured the rotation curves for the luminous matter of many spiral galaxies together with the observations of Bosma [19], who compiled 21 cm rotation curves for neutral hydrogen gas that extended far beyond the luminous matter of each galaxy, showed that the composite rotation curves were essentially "flat" out to the edge of the 21 cm data. This implied that if Newton's law of gravity was approximately valid, as in the Solar System, considerably more mass was required to be present in each galaxy. This invisible matter was called *dark matter*.

This led to the introduction of the "dark matter hypothesis" by Ostriker et al. [20], who concluded that the rotation curves of spiral galaxies could most plausibly be understood if the spiral galaxy was embedded in a giant spherical halo of invisible "dark matter" that provided a large contribution to the gravitational field at large distances from the center of the galaxy.

This dark matter hypothesis is very dubious, since to date no dark matter has been definitely *detected* and the nature of dark matter remains *unknown* [21].

The notion of "dark energy" arose from two sets of observations [22, 23] that suggested that the expansion of the universe is *accelerating*. These observations were very surprising and unexpected, since it was generally considered that the spatial expansion of the universe should be slowing down due to the gravitational attraction of the galaxies.

Both sets of observations were based upon the analysis of supernovae of Type Ia, which are considered to be excellent standard candles across cosmological distances, and allow the expansion history of the universe to be measured by considering the relationship between the distance to an object and its redshift, which indicates how fast the supernova is receding from us. Both teams found that the supernovae observed about halfway across the observable universe (6–7 billion light-years away) were dimmer than expected and concluded that the expansion of the universe was accelerating rather than slowing down as expected.

The conclusion from this observation was that the universe had to contain enough energy to overcome gravity. This energy was named "dark energy." The amount of dark energy in the universe, assuming the validity of the SMC, is estimated to be about 68% of the total mass-energy existing in the universe.

According to Peebles and Ratra [24], dark energy is a hypothetical form of energy that pervades the whole of space and causes the expansion of the universe to accelerate at large cosmological distances. Currently there exists no accepted physical theory of dark energy, suggesting that the existence of such energy is a dubious assumption of the SMC.

3. Discussion and conclusion

This book is divided into two main sections. Section 2 is devoted primarily to alternatives to the Big Bang scenario of the SMC based upon modifications of the Steady-State Models that were popular prior to 1965. Section 3 contains chapters that discuss modifications to the GTR.

Chapter 2 considers the "Tired Light" hypothesis introduced by Zwicky in 1929, in which the redshift is assumed to occur by the photons loosing energy due to interactions with material particles as they travel through cosmological space. The authors indicate that this assumption satisfies many of the observations and overcomes some of the problems of the Big Bang hypothesis.

Chapter 3 presents a different Steady-State Model as another alternative to the Big Bang hypothesis. The author discusses the possible existence of repulsive electromagnetic force fields emanating from galactic super-massive black hole cores that cause the expansion of the universe, although purely gravitational dynamics is maintained within each galaxy. The author indicates the implication of these electromagnetic fields upon the SMC.

Chapter 4 describes the effects of the large scale magnetic fields observed in galaxies and clusters of galaxies upon the Big Bang scenario and the CMB. The authors discuss the origin of such large scale magnetic fields and in particular, analyze their effects upon the CMB anisotropies.

Chapter 5 discusses alternative models for gravitational interactions that provide a generalization of the field equations of the GTR. In particular, it reviews cosmological models based upon a "polynomial affine gravity" scenario and discusses in some detail several cosmological solutions, especially those in the relativistic limit, in which the torsion vanishes.

Chapter 6 proposes a generalization of the Equivalence Principle for quantum gravity to reconcile quantum mechanics and general relativity. It defines the "Equivalence Principle of quantum gravity" to be "The laws of physics must be of such a nature that they apply to systems of reference in any kind of motion, both classical and quantum."

In conclusion, I believe that the dubious assumptions of the SMC will only be overcome when both the incompatible SMPP and the GTR are replaced by a quantum gravitational field, especially one based upon a particle physics model that has the appropriate properties to provide an understanding of both dark matter and dark energy [25].

Author details

Brian Albert Robson
The Australian National University, Canberra, Australia

*Address all correspondence to: brian.robson@anu.edu.au

References

[1] Ade PAR et al. (Planck Collaboration). Planck 2013 results. I Overview of products and scientific results. Astronomy and Astrophysics. 2014;**571**:A1, 48pp

[2] Gottfried K, Weisskopf VF. Concepts of Particle Physics. Vol. 1. New York: Oxford University Press; 1984. 189pp

[3] Einstein A. The basics of general relativity theory. Annals of Physics. 1916;**49**:769-822

[4] Robson BA. The generation model of particle physics. In: Kennedy E, editor. Particle Physics. Rijeka: InTech; 2012. pp. 1-28

[5] Friedmann A. Über die Krümmung des Raumes. Zeitschrift für Physik A. 1922;**10**:377-386

[6] Lemaître G. Un Univers homogène de masse constante et de rayon croissant rendant compte de la vitesse radiale des nébulueses extra-galactiques. Annales de la Societe Scientifique de Bruxelles. 1927;**47**:49-59

[7] Robertson HP. Kinematics and world structure. The Astrophysical Journal. 1935;**82**:284-301

[8] Walker AG. On Milne's theory of world structure. Proceedings of the London Mathematical Society. 1937;**42**: 90-127

[9] Hubble E. A relation between distance and radial velocity among extra-galactic nebulae. Proceedings of the National Academy of Sciences. 1929; **15**:168-173

[10] Gamow G. The origin of elements and the separation of galaxies. Physical Review. 1948;**74**:505-506

[11] Alpher RA, Herman R. Evolution of the universe. Nature. 1948;**162**:774-775

[12] Smoot G, Davidson KD. Wrinkles in Time. London: Little, Brown and Company (UK) Ltd; 1993. 315pp

[13] Guth AH. Inflationary universe—A possible solution to the horizon and flatness problems. Physical Review D. 1981;**23**:347-356

[14] Narlikar JV, Wickramasinghe NC. Microwave background in a steady state universe. Nature. 1967;**216**:43-44

[15] Bondi H, Gold T. The steady-state theory of the expanding universe. Monthly Notices of the Royal Astronomical Society. 1948;**108**:252

[16] Hoyle F. A new model for the expanding universe. Monthly Notices of the Royal Astronomical Society. 1948; **108**:372

[17] Penzias AA, Wilson RW. A measurement of excess antenna temperature at 4080 Mc/s. The Astrophysical Journal. 1965;**142**: 419-421

[18] Rubin VC, Ford KW, Thonnard N. Rotational properties of 21 Sc galaxies with a large range of luminosities and radii, from NGC 4605 (R = 4 kpc) to UGC 2885 (R = 122 kpc). The Astrophysical Journal. 1980;**238**:471-487

[19] Bosma A. The distribution and kinematics of neutral hydrogen in spiral galaxies of various morphological types [thesis]. Groningen University; 1978

[20] Ostriker JP, Peebles PJE, Yahil A. The size and masses of galaxies and the mass of the universe. The Astrophysical Journal. 1974;**193**:L1-L4

[21] Robson BA. The story of dark matter. In: Capistrano de Souza AJ, editor. Essentials on Dark Matter. Rijeka: InTech; 2018. pp. 9-22

[22] Riess AG et al. Observational evidence from supernovae for an accelerating universe and a cosmological constant. The Astronomical Journal. 1998;**116**:1009-1038

[23] Perlmutter S et al. Measurements of omega and lambda from 42 high-redshift supernovae. The Astrophysical Journal. 1999;**517**:565-586

[24] Peebles PJE, Ratra B. The cosmological constant and dark energy. Reviews of Modern Physics. 2003;**75**: 559-606

[25] Robson BA. Cosmological consequencies of a quantum theory of mass and gravity. In: Capistrano de Souza AJ, editor. Trends in Modern Cosmology. Rijeka: InTech; 2017. pp. 159-176

Section 2

Alternatives to the Big Bang of the SMC

Chapter 2

Tired Light Denies the Big Bang

Ming-Hui Shao, Na Wang and Zhi-Fu Gao

Abstract

More and more problems related to Big Bang have been appeared in recent years. All the problems are due to the Doppler interpretation of redshift. The "tired light" theory, proposed in 1929 by Zwicky and most recently developed by Shao in 2013, gives a new explanation for redshift. The theory has shown that the redshift is induced from the energy loss of photons by the interaction with material particles on their journey through cosmological space. The basic principles related to the energy transfer are mainly the mass-energy equivalence and the Lorentz theory. Problems, such as super velocity, the horizon problem, the cosmological microwave background radiation, and Olbers' paradox, vanish in the cosmological model of "tired light" theory. The model describes a boundless and timeless Cosmos.

Keywords: tired light, energy loss, photons, cosmological redshift, Lorentz theory, big bang, CMBR, Olbers' paradox

1. Introduction

Cosmology is as old as other branches of sciences, beginning at the ancient Greeks. But modern cosmological study started in the twentieth century, marked by Einstein's theoretical research in 1917 and Hubble's observational investigations in 1929. The Big Bang cosmological model came mainly from Hubble's work. Hubble used the Doppler Effect to interpret what came to be known as the cosmological redshift. The "tired light" hypothesis was proposed by Zwicky in 1929, after Hubble's paper, as an alternative interpretation to that of the Doppler effect for the cosmological redshift [1]. In 1929, Hubble obtained a distance-redshift relation through observations. He then obtained a new relation of distance-velocity by using the Doppler effect to interpret the redshift. About half a year later after Hubble's paper, Zwicky proposed a "tired light" hypothesis to explain the distance-redshift relation. But the nature of the "tired light" was only vaguely explained in Zwicky's work, so that the "tired light" hypothesis has not been accepted by most cosmologists and astronomers to this day. The Big Bang, after Hubble's work, became the most accepted cosmological model. In recent years, problems related to Big Bang have been more and more clearly realized by cosmologists and astronomers. Some problems are directly related to the interpretation of the Doppler effect for cosmological redshift. The Big Bang model cannot surmount these problems. Fortunately, the study of "tired light" theory has continued. In 2013, Shao developed the "tired light" hypothesis on the basis of physical principles, that is, (a) electromagnetic field theory, (b) the mass-energy equivalence, (c) the quantum light theory, and (d) the Lorentz theory [2]. Based on these physical principles, the "tired light" theory explains the cosmological redshift as the result of photon energy loss due to the interactions with material particles as photons travel through cosmological space. By this interpretation for cosmological redshift, the Cosmos is infinite and eternal.

2. Big Bang, history and problems

Hubble derived the distance-velocity relation from observational result of the distance-redshift relation, by employing the interpretation of Doppler effect [3]. The Big Bang is popularly known in present days, but some problems accompanying it have been aroused. Furthermore, some problems result from the interpretation of Doppler effect for the cosmological redshift. Actually, all the problems are rooted in the Doppler effect interpretation of cosmological redshift. The present situation is that the Big Bang cosmology is facing some hurdles, which it seems cannot be easily overcome within the framework of the Big Bang model.

2.1 Big Bang and Doppler effect

The Big Bang model came from two sources. One source is, weakly, Einstein's finite boundless cosmological model proposed in 1917. The other one is, strongly, the interpretation of the Doppler effect for the cosmological redshift, employed by Hubble [3].

Hubble employed the Doppler effect to interpret the cosmological redshift in the distance-redshift relation he discovered in 1929. In doing this, Hubble derived the distance-velocity relation which led people to conceive of the Cosmos in the image of a Big Bang. Why did Hubble use the Doppler effect to interpret the redshift? One reason is that he had no other choice since the Doppler effect was the only interpretation for redshift at that time.

Reber had introduced the history of the Doppler effect and its application to the studies of the Sun's motion and the rotation of our galaxy [4]. The Doppler effect was enunciated in 1842. Doppler claimed that the frequency and wavelength of light or sound would change when a signal from a moving source was observed. The effect was confirmed in 1845 for sound. It was subsequently confirmed for light by observation in 1871 and by experiment in 1901.

About the year 1900, the Doppler effect was used to study the rotating of double stars. Around 1910, it was used to study the motion of the Sun in the Milky Way. And by 1920, it was used to examine the rotation of our galaxy. "All three of the above examples are correct interpretations of spectral shift caused by relative motion between the source and the observer," Reber remarked.

When Hubble had found the relation of distance-redshift, he used the Doppler effect to interpret the redshift, that is, the movements of spectral lines. He did so habitually, as previous studies had done. Whether he considered the difference between light sources within our galaxy and those in other galaxies is not known. But the problems were brewing by his doing. Hubble should have been conscious of the fact that the light sources belong to other galaxies. Then, was it suitable to use the Doppler effect for the interpretation of the redshift in the distance-redshift relation? Nevertheless, Hubble transformed the distance-redshift relation to the distance-velocity relation by using a Doppler interpretation for the redshift. In doing this, Hubble had no real choice because there was only the Doppler effect available to him for the interpretation of redshift. Regarding this, Reber remarked: "clearly, the interpretation of these spectral shifts as representing relative motion was dubious."

2.2 The problems related to Big Bang

As a cosmological model, the Big Bang presents some difficult problems. There are some phenomena that the Doppler effect cannot explain and others that Big Bang cannot resolve, due to its Doppler effect interpretation for cosmological redshift.

Some of the phenomena and problems are:

1. The solar limb effect, that is, the variation of redshift on solar disc,
2. The signal redshift of Pioneer 6,
3. The large redshift of quasars,
4. Super velocity of light,
5. The horizon problem,
6. The age of the Cosmos.

3. The history of "tired light" theory

Hubble employed the Doppler effect to interpret the cosmological redshift in his study. It was a bold move which he might not have made if he had considered the issue deeply. Zwicky however did not think the Doppler effect was suitable for the interpretation of the cosmological redshift. The Doppler effect says nothing about the nature of matter. It is only a problem in kinematics. The redshift induced by the Doppler effect is caused by relative motion between the light source and the observer. Zwicky thought the things are not so simple. He proposed the "tired light" hypothesis that the cosmological redshift is caused by the interaction of the light with a latent feature of the Cosmos. The tired light hypothesis claims that while the light propagates, it must be affected by all matters of the Cosmos. The idea came from Mach who thought that all of the matter in the Cosmos is related, so that any part of the matter is affected by all the other parts. Although Zwicky objected to the Doppler effect interpretation of cosmological redshift, the tired light hypothesis was vague on physical mechanics, so few people took the hypothesis seriously. But there have been a few people contemplating tired light, keenly working to find the physical mechanics thereof, without success [5–11]. The physical mechanics of tired light was not clearly described until 2013 when M. Shao published his paper. Shao pointed out that the physical mechanics of tired light should be the Lorentz electric force produced by the electromagnetic field of photons acting on material particles.

The phenomena and problems related to Big Bang listed in Section 2.2 can now be explained by the renewed tired light theory (thereafter referred as TLT).

4. The TLT: basic thoughts

The Cosmos is composed of matters and fields. A material aggregation produces two kinds of field, the gravitational field and the electromagnetic field, with the forces of the fields; the gravitational and the electromagnetic forces; and different matter aggregations interact with one another. They attract for or repel each other, which changes or keeps the conditions of matter distribution of various regions, in large or small scales, of the Cosmos.

With the principle of matter-energy equivalence, the electromagnetic field and material particles can be considered the same thing. Then, their densities and sizes can be calculated. A hydrogen atom and a photon of visible light are presented in the simplest case.

The mass of a hydrogen atom is 1.67×10^{-27} kg, and its diameter is 2.4×10^{-10}m. Then, the density of the hydrogen atom is 231 kgm^{-3}.

The average wavelength of the visible light is 5.5×10^{-7} m, being the diameter of a photon. For the simplicity, we assume that the photon has spherical symmetry. From $E = h\nu$, here h is the Plank constant, and ν is the frequency of the light, and $E = Mc^2$, the mass of the photon is 4×10^{-36} kg. Then, the density of the photon is $4.6 \times 10^{-17}\mathrm{kgm}^{-3}$.

From the above data, the ratio of the size of the photon to that of the hydrogen atom is 2292, and the ratio of the density of the photon to that of the hydrogen atom is 2×10^{-19}. The two ratios reveal that the hydrogen atom is very hard and small compared to the photon. The photon, relatively, is extremely low in density and more than 2000 times larger than the hydrogen atom. We now consider a hydrogen atom encountering a photon. The photon is traveling at a light speed, 3×10^8 ms^{-1}, in cosmological space. The hydrogen atom is more or less stable. Since the motion is relative between the photon and the hydrogen atom, the photon could be supposed in a stable mode, and the hydrogen atom penetrates the photon at the speed of light. The diameter of the photon being 5.5×10^{-7} m, then the hydrogen atom should penetrate the photon in 1.8×10^{-15} s. In such a short time, the hydrogen atom should not show electronic neutrality but present in the mode of electric dipole. A photon is actually a section of moving electromagnetic field. During the time of hydrogen atom penetration of the photon, the electromagnetic field of the photon should interact with the hydrogen atom. The electromagnetic field of the photon acts on the hydrogen atom and does some work on the hydrogen atom, given by the Lorentz electric force. A little bit of the energy of the electromagnetic field is transferred from the electromagnetic field to the hydrogen atom. Although the photon loses a very small amount of energy in meeting with a hydrogen atom, it will, traveling a long distance in the cosmological space, show a detectable effect for a photon meeting a large number of hydrogen atoms (and other kinds of atoms and molecules), that is, the photon undergoes a cosmological redshift.

The expression for the cosmological redshift, based on the tired light theory, obtained by Shao in 2013, is, $Z = \exp(kN\lambda_0 + u) - 1$, where k is a coefficient, N is the number of material particles that a photon interacts with in its course from emitter to observer, λ_0 is the original wavelength of the light, and u is the change of the wavelength induced by the gravitational effect [2].

5. Reasoning: the energy loss of a photon, the electromagnetic field of a photon, etc.

5.1 Polarization of atoms and molecules

A hydrogen atom is electrically neutrality in a common time scale. But in a very short instant, for example, about the period of an electron revolving around the nucleus, the hydrogen atom appears to be polarized. A polarized atom must be affected by an electromagnetic field. The hydrogen atom is the simplest atom. It is believed that many kinds of atoms are similarly polarized in a very short time interval.

A small molecule is not always oscillating, deviating from the mode of electromagnetic equilibration. Hence, in a very short time interval, a molecule may be affected by an electromagnetic field. Based on this reasoning, it may be said that many kinds of atoms and molecules should be affected by electromagnetic fields.

5.2 The electromagnetic field of a photon

Light has been considered from two different viewpoints, the wave theory and the particle theory. To explain the photoelectron effect, Einstein suggested that light energy propagates in packets, that is, photons. This is the particle viewpoint of light. The energy of a photon is $E = h\nu$. Contrarily, from the viewpoint of wave theory, light is looked upon as a propagating series of electromagnetic wave.

The two different viewpoints describe the same objective thing but they express different features of light. For the purpose of understanding the energy transfer process from photons to material particles, both viewpoints need to be combined. What is a photon? Einstein and Plank defined a photon by the frequency of the light. They were talking about the effects of energy emission and absorption. For the special case under discussion here, a photon is looked upon as a section of the electromagnetic field of the traveling light. The length of the section is the wavelength of the light. In the corresponding period, the electromagnetic field completes an oscillation. A thread of traveling light can be imaged as a series of photons, and a photon is a section of moving electromagnetic field.

Consider the case of a photon interacting with a hydrogen atom. Since the average size of a photon of visible light is more than 2000 times larger than a hydrogen atom, as mentioned above, the situation of a photon interacting with a hydrogen atom can be viewed as the hydrogen atom penetrating the photon. The photon is viewed as a stable electromagnetic field, and the hydrogen atom is moving through the electromagnetic field of the photon.

5.3 The Lorentz force of the electromagnetic field of a photon

The size of the electromagnetic field is the wavelength of the light. That is to say that a photon is a section of electromagnetic field which, in a period, oscillates along its length. Image moving together with an electromagnetic field is within the cosmological space. We will see some hydrogen atoms passing through the electromagnetic field, that is, they are penetrating the photon.

The electromagnetic field of the photon will affect a charged particle with a Lorentz force. In the short time of 1.8×10^{-15} second, the hydrogen atom is polarized, and it should be forced to change its motion mode by the Lorentz force. The Lorentz magnetic force changes the moving direction of the hydrogen atom, and the Lorentz electric force does some work on the hydrogen atom. Thus, some energy, although small, is transferred from the photon to the hydrogen atom. In the tremendous long journey from emitter to receiver, a photon has to encounter a large number of material particles, so the sum of the energy loss of the photon should show up. The energy loss of the photon is the redshift of the light.

5.4 The ISM and IGM

The Cosmos is the only objective existence and is composed of all the matter and fields produced by the matter. Since it is the only existent thing, it is eternal and infinite. Since it is eternal and infinite, matter can exist in all possible forms, of which there are galaxies, stars, planetary systems, planets and others. All the above are aggregations of atoms, except that there are roaming atoms, molecules, and dusts, which are the components of the interstellar medium (ISM) and intergalactic medium (IGM). By the way, since the Cosmos possesses infinite possibilities for matter, all kinds of living things, including the human, are here on Earth.

5.5 Matter and fields

There are four kinds of forces related to matter. The strong force and the weak force, which manifest within atoms, are not discussed here. The electromagnetic force and the gravitation are the forces controlling the material distribution of the Cosmos. Generally speaking, all forms of material aggregation and the two kinds of field, the electromagnetic field and the gravitational field, are the elements composing the Cosmos. Matter forms galaxies, stars and planets, and still more smaller celestial objects. The concern here is the atoms and molecules which roam in the interstellar space and intergalactic space, that is, the ISM and the IGM. The greater part of the ISM and IGM is composed of the simplest atom, hydrogen. Kant first talked about the formation process of planetary system. Generally, he is right, and, his reasoning is applicable to the formation of galaxies. These processes are condensing processes. The question is what is the inverse process by which matter is exiled? Roughly, the supernova is a means of material dispersion. Another way is the evaporation of black holes. The Cosmos is in an equilibrium state by these two processes: condensing and dispersing. Nevertheless, we are especially interested here in the interaction between photons and material particles, atoms and molecules.

Most of the ISM and IGM are hydrogen atoms, and most of the others are helium. The rest are heavier elements, molecules, etc. As discussed previously, when a photon meets a material particle, the photon can be looked upon as a section of moving electromagnetic field, and the atom or molecule is passing through the electromagnetic field. While the atom or molecule travels through the electromagnetic field of the photon, a little bit of energy is transferred from the photon to the particle through the effect of the Lorentz electric force. Within the tremendously long journey of a photon from emitter to observer, a vast number of atoms and molecules are encountered and interacted with the photon. With the interactions, a part of the energy of the photon is consumed, and the photon is redshifted.

5.6 Equilibrium of energy matter and energy transfer

As a whole, the cosmos is in an equilibrium state of the matter and energy. But in a local area, there are stars forming and extinguishing. On the larger scale, there are galaxies forming and extinguishing (dispersing). All the processes relate to energy absorption or emission. Consider again the energy transfer process from a photon to a material particle. As said before, when a photon meets a material particle, it can be looked upon as a material particle moving through an electromagnetic field. The material particle may be an atom or a molecule. The particle may be charged. If not, the electrically neutral particle may display polarity in the very short interval as it moves through the electromagnetic field of the photon. When the charged or polarized material particle moves through the electromagnetic field, the Lorentz magnetic force of the electromagnetic field changes the direction of the path of the material particle, and the Lorentz electric force does some work on the material particle, changing the motion of the particle. Then some energy is transferred from the electromagnetic field, that is, the photon, to the material particle. In the long journey of the photon from emitter to observer, a massive number of material particles have interacted with the photon, producing an observable effect, that is, the redshift.

6. The equation of the cosmological redshift

6.1 The energy loss of a photon

A larger photon will transfer more energy to a material particle since it has a longer interaction time with the material particle. So a larger photon transfers a larger part of its energy to the material particle than a smaller one. Thus, the energy loss of a photon is proportional to its size. A photon, on its journey after emission, meets a number of material particles before being received by an observer. The greater the number of material particles it meets, the more energy it loses. So, the energy loss of the photon is also proportional to the number of material particles it meets.

6.2 Equations for the cosmological redshift

When a photon of size λ and energy E meets a material particle, the material particle runs through the electromagnetic field of the photon in a time $t = \lambda/c$, where c is the speed of light. In the interaction, the material particle can be viewed as stationary compared to the speed of the photon. During the interaction, the photon transfers a tiny amount of energy $\delta(E)$ to the material particle. A coefficient $k = \delta(E)/E\lambda$ is defined here, denoting the rate of energy loss of the photon per unit length. The coefficient k is denoted conceptually at this stage. Further theoretical or experimental studies are needed to determine its value.

If a photon of size λ_0 and energy E when emitted meets N material particles in its path and transfers a part of its energy to the material particles, supposing all the material particles interact equally with the photon, a differential equation for the energy of the photon is obtained with coefficient k as follows:

$$\frac{dE}{dN} = -k\lambda_0 E. \tag{1}$$

The solution, from the condition $E = E_0$ when $N = 0$, is

$$E = \frac{E_0}{\exp(kN\lambda_0)} \tag{2}$$

The energy loss of the photon is $\Delta E = E_0 - E$. Thus, there is

$$\Delta E = E_0\left(1 - \frac{1}{\exp(kN\lambda_0)}\right). \tag{3}$$

The expression for the redshift is $Z = \frac{\lambda - \lambda_0}{\lambda_0}$. It can be written as $Z = \frac{\nu_0 - \nu}{\nu} = \frac{E_0 - E}{E}$. Then, it obtains,

$$Z = \frac{\Delta E}{E}. \tag{4}$$

From Eqs. (2)–(4),

$$Z = \exp(kN\lambda_0) - 1. \tag{5}$$

6.3 The gravitational redshift

The redshift described above is induced by the process of energy loss of photons. It can be called tired light redshift, referred to as "TR" thereafter. The cosmological redshift (CR) is not simply induced by a single effect. In addition to TR, the gravitational redshift (GR), that is, the redshift induced by the gravitational field of the source star, must be considered in the analysis and evaluation of CR.

The GR is different from TR in nature. It has no relation to the interaction of photons with material particles. So, Eq. (5) should have the form

$$Z = \exp\left(kN\lambda_0 + \lambda_g\right) - 1, \tag{6}$$

where λ_g denotes the part of the wavelength change induced by the gravitational effect. (GR equals GM/c^2R. Here, it is also expressed as λ_g/λ_0.) For simplicity, set $u = \lambda_g$. Eq. (6) is rewritten as,

$$Z = \exp\left(kN\lambda_0 + u\right) - 1. \tag{7}$$

Now, there are two expressions, Eqs. (5) and (7), for the cosmological redshift. Eq. (5), comparatively simpler, considers only the effect of TR, whereas Eq. (7) considers the effects of TR and GR.

7. Features of TLT and evidence: part I

7.1 Redshift vs. wavelength

A photon emitted from a star undergoes a continuous process of energy loss on its journey by interacting with material particles before it reaches an observer on Earth. It encounters material particles within the corresponding spaces of: (1) the atmosphere around the star, (2) the ISM in the galaxy the star belongs to, (3) the IGM, (4) the ISM of our galaxy, and (5) the atmosphere of the Earth. In the five parts, the IGM is the main one which CR is induced by. Although, the IGMs are sparsely distributed in the intergalactic space, the space is vast compared with the other spaces. Therefore, the tired light redshift (TR), the main part of the cosmological redshift (CR), is mainly induced by the interaction of photons with material particles of IGM. It may then be supposed that the photons of different wavelengths emitted from a certain source meet the same number of material particles in the intergalactic space along the line of sight of an observer on the Earth. Furthermore, roughly speaking, it may be supposed that the photons meet the same number of material particles on their entire journey from an emitter to an observer. So, N, the number of material particles the photons met can be considered as a constant. Thus, N can be included in the coefficient k, and Eq. (7) takes the form,

$$Z = \exp\left(k\lambda_0 + u\right) - 1. \tag{8}$$

The first-order approximation to Eq. (8) is

$$Z = k\lambda_0 + u. \tag{9}$$

Eq. (7) shows the characters that a larger λ_0 is related to a larger Z, and a larger N is related to a larger Z too.

7.2 Evidence for TLT

7.2.1 Early evidence

As early as 1929, Zwicky had noticed the relation between redshift and wavelength. "Some exceptions have been found, suggesting that $\Delta\nu/\nu$ for H_β is somewhat greater than for H_γ" [1]. Here, $\Delta\nu/\nu = \Delta\lambda/\lambda_0$ is the redshift, and the value of λ_0 for H_β is greater than that for H_γ. This is the earliest description for the relation between redshift and wavelength, whereby a longer wavelength is related to a larger redshift. Eq. (7) shows this character; a larger λ_0 is corresponding to a larger Z.

Wilson reported an observational result for the Seyfert galaxy NGC4151 [12]. He noticed a "slight apparent trend of velocity with wavelength." In that study, redshift is interpreted in terms of the Doppler effect and expressed as a recession velocity, that is, $V = cZ$. Then, from Eq. (9),

$$V = k'\lambda_0 + u'. \tag{10}$$

The fitting result for the relation between the velocity V and wavelength λ_0 to Eq. (10) in Eq. (2) is,

$$V = 0.003959\lambda_0 + 948.09. \tag{11}$$

Comparatively, the mean radial velocity obtained by Wilson is 967 km s^{-1}.

Espey et al. presented a set of data for redshifts in a range of about 1–3, of six emission lines from 18 quasars, with mean values of velocity for five lines relative to H_α [13]. The fitted result for the data is,

$$V_{relative\ to\ H_\alpha} = 0.21739\lambda_0 - 1275.8. \tag{12}$$

7.2.2 More evidence

Schmidt and Matthews presented observational results of emission lines for quasars 3C47 and 3C147. Seven lines were observed for 3C47 and five lines for 3C147 [14]. After necessary treatment for the data, relations between redshift Z and wavelength λ_0 have been fitted to Eq. (8) as follows [2]. For 3C47, the relation is,

$$Z = \exp\left(2.67 \times 10^{-7}\lambda_0 + 0.353\right) - 1. \tag{13}$$

For 3C147, the relation is,

$$Z = \exp\left(7.11 \times 10^{-7}\lambda_0 + 0.432\right) - 1. \tag{14}$$

Nishihara et al. presented redshifts of the emission lines H_α, H_β, OIII, M_gII, CIII, CIV, and OI for five quasars, Q1634 + 706, Q1630 + 377, Q0117 + 213, Q1011 + 250, and Q1331 + 170 [15]. The relations between redshift Z and wavelength λ_0 to Eq. (8) for each quasar in Eq. (2) are:

$$\text{For Q1634} + 706, Z = \exp\left(5.33 \times 10^{-7}\lambda_0 + 0.847\right) - 1, \tag{15}$$

$$\text{For Q1630} + 377, Z = \exp\left(6.41 \times 10^{-7}\lambda_0 + 0.904\right) - 1, \tag{16}$$

$$\text{For Q0117} + 213, Z = \exp\left(8.84 \times 10^{-7}\lambda_0 + 0.912\right) - 1, \tag{17}$$

$$\text{For Q1011} + 250, Z = \exp\left(3.64 \times 10^{-7}\lambda_0 + 0.968\right) - 1, \quad (18)$$

$$\text{For Q1331} + 170, Z = \exp\left(9.18 \times 10^{-7}\lambda_0 + 1.126\right) - 1. \quad (19)$$

7.2.3 The data of redshifts

In an ordinary redshift observation, usually some emission lines or absorption lines are detected. In most cases, the values of redshift of the lines are slightly different from each other. Early observers had published the original results. But afterwards, the deviations between the redshifts of the lines in most cases were moved out by averaging the values of redshift, the reason being that the redshifts for lines from a certain source should be the same according to the Big Bang model. Hence, all that is required is an average value. Subsequently, most observers gave only the average value of redshifts and did not mention the deviations, as if they do not exist, since they cannot be explained by Big Bang. But the difference between redshifts for lines from a certain source reveals a flaw in the Big Bang model. Nishihara et al. have remarked: "however, the physical mechanisms producing these velocity deviations are not well understood" [15].

The cosmological model of Big Bang is mainly inferred from the Doppler interpretation to CR following Hubble's lead. The main reason of Hubble used it is that the Doppler effect was the only interpretation for redshift at that time. Zwicky advanced an alternative, that is, the tired light hypothesis, which led to TLT (tired light theory) [1, 2]. TLT developed the tired light hypothesis on the foundations of physics, that is, electromagnetic field theory and the Lorentz force. It revealed the redshift-wavelength relation, substantiated by observational results as shown above.

8. Features of TLT and evidence: part II

8.1 Redshift vs. the number of material particles

For a given λ_0 included in the coefficient k, Eq. (7) becomes,

$$Z = \exp\left(kN + u\right) - 1, \quad (20)$$

showing the relation between redshift and the number of material particles a photon interacts with on its journey. If the material particles are assumed to be distributed evenly in intergalactic space or the other respective spaces, the redshift should be proportional to the distance of the photon's journey. Some redshift phenomena that cannot be explained by the Doppler effect can be explained by TLT.

The Limber effect of the Sun, the signal redshift of Pioneer 6, and the large redshift of quasars are the examples of some overt phenomena that cannot be explained by the Doppler effect. Eq. (11) explicitly shows the relation of the redshift to the number of material particles by which the three foregoing puzzles can be accounted for.

8.2 The limb effect (variation of redshift on the solar disc)

8.2.1 The Cosmos, Sun, Earth, and human beings

The Sun is a special star for human beings. It is the only star we can observe in detail since it is near the Earth. Actually, the Earth and we human beings and all the living things on the Earth are entwined with the Sun. We are a part of the solar system. We belong to the solar system, which belongs to our galaxy, which in turn belongs to the Cosmos. We human beings are simply a particular form of matter in the Cosmos. We observe the Cosmos and try to understand it with our peculiar intelligence that is self-understanding of the Cosmos from a viewpoint of philosophical significance.

8.2.2 Limb effect

The limb effect is a phenomenon involving redshift on the solar disc, that is, the redshift changing from the center to the limb of the solar disc. On the edge of the solar disc, the redshift is larger than that near the center. Although the limb effect was discovered more than a century ago, it could not be adequately explained [3]. Assis discussed the limb effect and concluded that the tired light theory provides a satisfactory explanation. He suggested that the redshift was due to the interaction of light with the atmosphere of the Sun while passing through it [16].

TLT gives a clear explanation of the Limber effect—the largest redshift on the limb of the solar disc is due to the fact that the light emitted from the edge of the surface of the Sun encounters more material particles while traveling to Earth and therefore loses more energy than that from the inner part of the solar disc, since the atmosphere of the Sun at the edge of the solar disc has the deepest length along the line of sight of an observer on Earth [2].

Adam, as early as in 1948, observed the redshift on the solar disc and presented a set of 14 redshifted lines at seven positions on solar disc. The redshifts vary from the center to the limb [17]. In 1991, LoPresto et al. observed the infrared oxygen triplet absorption lines at seven positions on the solar disc along the limb effect [18].

According to Eq. (11), redshift is related to the number of material particles that the photons met on their journey. Considering the depth of the atmosphere of the Sun along the line of sight of an observer on the Earth, Shao showed that Adam's data fit Eq. (11). The redshift curve coincides with the data points satisfactorily. Similarly, the data from LoPresto et al. were found to fit Eq. (11) very well too [2].

8.3 The signal redshift of Pioneer 6

When Pioneer 6 on its orbit at the far side of the Sun was approaching the Sun in November 1968, the signal it sent to Earth gave an additional frequency shift, or redshifted. The shift changed day by day. The phenomena could not be well explained. Chastel and Heyvaerts introduced the frequency shift [19]. Merat et al. reported that the data "strongly favor the existence of a new redshift cause at work in the Sun's vicinity" [20].

The signal redshift of Pioneer 6 can be explained by TLT. Like the explanation to the limb effect of the Sun, the signal redshift of Pioneer 6 is due to the energy loss in the signal while traveling through the atmosphere of the Sun. The atmosphere of the Sun is at a certain depth around the Sun. While the signal path from Pioneer 6 to Earth was getting closer to the edge of the Sun, the signal passed through a longer distance, day by day, through the atmosphere of the Sun. Consequently, more energy was lost and so the stronger the redshift caused, day by day, until Pioneer 6 went behind the Sun.

8.4 The large redshift of the quasars

The large redshifts of the quasars are rather queer, as their nature is not clear. TLT may give some insight into the quasars. The quasars might have much thicker and denser layers of atmosphere, that is, gaseous material particles, compared to normal stars. On this assumption, TLT provides a possible explanation for the large redshifts of the quasars. The main part of CR is TR, which is induced by the interaction of photons with material particles. The greater the number of material particles that the photons encounter the larger the redshift that should result. The light emitted from a quasar has to penetrate through the dense and thick layer of atmosphere around it, so that the light is redshifted much more than that from a normal star at the same distance from Earth. Thus, the quasars do not need to be located so far away as the Doppler effect interpretation of redshift supposes.

9. Considerations about the problems related to Big Bang

9.1 The problems of Big Bang

Hubble's work had led to the Big Bang model by using the Doppler effect to interpret the cosmological redshift. Of the two possible alternatives to Hubble for the observed redshift, either employing the Doppler interpretation or giving no interpretation, Hubble selected the former, to some extent beyond the traditional usage of the Doppler effect. In so doing, Hubble had triggered the Big Bang, although he harbored doubts as to its legitimacy. In 1937, he remarked: "thus the familiar interpretation of redshift as velocity shifts leads to strange and dubious conclusions." In contrast, as to the tired light interpretation for redshift, Hubble remarked, "while the unknown, alternative interpretation leads to conclusions that seem plausible and even familiar" [21].

Because the Big Bang is rooted in the Doppler interpretation for cosmological redshift, many problems are produced thereof, puzzling cosmologists to this day. These problems are insoluble by Big Bang cosmology because they are largely characteristic of the Doppler effect, from which researchers cannot exculpate themselves. Among the problems loom the super velocity problem, the horizon effect, and most importantly, the problem of the beginning of the Cosmos. There are other problems also related to the cosmological model of Big Bang, famously the cosmic microwave background radiation (CMBR) and the old paradox of Olbers. Comparatively, all these problems do not arise in the tired light theory (TLT).

9.2 Super velocity

The super velocity problem is the direct consequence of invoking the Doppler interpretation of cosmological redshift to obtain $V = H_0D$. Recalling history, Hubble and Humason had adduced a distance-redshift relation (DRR) through observations. Following traditional lines, Hubble transformed DRR to a distance-velocity relation (DVR) by replacing redshift with velocity, from which it has been concluded that the galaxies are moving away from the Milky Way. Hubble accepted the idea of runaway galaxies as a fact. But is it really a fact? The answer must be "no." The supposed "fact" simply lends support to the Big Bang model, hence its raison d'être. In the treatment from DRR to DVR, there is no physical meaningful content about the nature of matter at all. Zwicky, therefore, objected and proposed his "tired light" hypothesis, which, afterwards, Hubble said is "plausible" and "familiar."

Zwicky's hypothesis has not been accepted by most astronomers and cosmologists because it is vague on physical meaning. But things are changing. It is especially important at present because more and more people have realized the problems related to Big Bang entanglement. After 84 years, the tired light hypothesis has been reset in accord with the foundations of physical principles [2]. The basic principles that TLT is based on are: (a) electromagnetic field theory, (b) the matter-energy equivalence, (c) the quantum theory of light, and (d) the Lorentz theory. According to TLT, since redshift is produced by energy transfer from photons to material particles, it concludes that there is no systematic motion of galaxies on large scale of the Cosmos. Hence, there is no super velocity problem. The Cosmos is boundless and timeless.

9.3 Horizon problem

The horizon problem (also known as the homogeneity problem) is a characteristic problem of the Big Bang model. It arises from the homogeneity of regions in the Cosmos, which cannot be explained by the Big Bang model. According to Big Bang, the history of the Cosmos is finite, and in a finite time, the Cosmos could not evolve to the present homogeneity. Because the Big Bang model is based on the Doppler interpretation for cosmological redshift, the problem is, then, inherited from the Doppler interpretation. If the Doppler interpretation of cosmological redshift is abandoned, the problem disappears. The cosmological model based on TLT does not produce the problem. Based on TLT, the Cosmos is infinite and eternal, so that the homogeneity of the Cosmos is natural.

9.4 The age of the Cosmos

The age of the Cosmos is another feature of the Big Bang model, again due to the Doppler interpretation of redshift. By using the Doppler effect for redshift, Hubble obtained the relation $V = H_0 D$. Since H_0, the Hubble constant, has the dimension v/d, then $1/H_0 = S$. Thus, S has been labeled the age of the Cosmos. But Hubble took the wrong direction when he interpreted the cosmological redshift by the Doppler effect. Everything derived from this wrong direction is also wrong. In the Big Bang model of the Cosmos, the cause is the result, and the result is the cause. Actually, the Doppler effect used in the interpretation of cosmological redshift and the Big Bang model describe the same thing, that is, recession of the celestial objects being observed. There is, between the two, nothing related to the material nature or physical process except kinematics. The TLT model of the Cosmos does not possess this problem.

9.5 The cosmic microwave background radiation (CMBR)

The CMBR was discovered in the 1960s, and it has been thought to be proof of the Big Bang. Just as Hubble had used the Doppler effect to interpret the redshift in the relation of redshift-distance because he had no alternative choice at his disposal, proponents of CMBR had no other explanation for it except Big Bang, at the time of the discovery. Now, the situation has changed. TLT interprets the redshift on a profound basis of physical principles and, at the same time, gives a plausible explanation for CMBR. The CMBR is tired light in the microwave band. The photons from all directions emitted by the faraway sources are redshifted after a long journey. Photons then, from all the other galaxies in the background of the Cosmos, around the Earth, theoretically around the Milky Way, have been redshifted to form the CMBR. Tired light does not only form CMBR, but it also forms CRBR (cosmological radio background radiation) [4].

9.6 Olbers' paradox

Olbers' paradox is a historical problem as old as natural science itself, that is, from the time of the ancient Greeks. Olbers described the paradox in 1823. After the Big Bang became a popular cosmological model, Olbers' paradox was explained by Big Bang because the history of the Cosmos was finite. But if Big Bang is not assured, then the explanation by it is not reliable. A new explanation could be given by the principle of TLT. By TLT, light from the stars in faraway galaxies should be redshifted such that visible light would be lengthened outside the waveband of visible light, then the night sky should be dark. As mentioned above, visible light has been redshifted into CMBR and CRBR. Since the energy lost by photons is transferred to material particles, the Cosmos may not be heavily heated. Material particles that have gained the energy have more ability to form new stars and galaxies.

10. Some thoughts on cosmology

Zwicky first proposed TLT because application of the Doppler effect to interpret cosmological redshift leads to a strange idea, that is, Big Bang. The etiology of the logical reasoning of TLT presented here is also uncomfortable feelings by the dizzy image of Big Bang.

We human beings have been living on the Earth for several million years. Our system of knowledge began in the Neolithic Age, not more than 10,000 years. The beginning of the sciences dates back to the age of ancient Greeks, about 2500 years ago. The beginning of modern sciences is marked by the work of Copernicus, not more than 500 years ago. As a branch of modern science, the history of modern cosmology is not more than 100 years. But the history of the knowledge of the Cosmos is as old as the beginning of science. On the one hand, cosmology is old since it began from and needs only reasoning on the whole nature. On the other hand, it is young because the detailed and elaborate observation methods and technology for the study of the Cosmos emerged at the start of the twentieth century. Thus, paradoxically, cosmology is the oldest and also the youngest branch of science.

Big Bang cosmology is only the beginning of modern cosmology. The tired light theory described herein, began with Zwicky's hypothesis, has been reset on the basis of physical principles. It may be the next forward step of modern cosmology.

11. Conclusions

The tired light theory, proposed by Zwicky in 1929 and recently developed by Shao in 2013, explains the cosmological redshift as the result of energy loss of photons due to the interactions with material particles as photons travel through cosmological space. A differential equation is established through the analysis of a photon's energy loss based on the mass-energy equivalence and the Lorentz theory. A redshift expression is achieved by expanding the solution of the equation. The redshift expression shows that the redshift is related to the wavelength of light and the number of material particles that photons interact with on their traveling journey in the cosmological space. The relationship of redshift to wavelength of light is in accordance with the observational data in the cited literatures. And the relationship of redshift to the number of material particles that interact with photons explains the limb effect of the Sun, the signal redshift of Pioneer 6, and the large redshifts of quasars. The cosmological model based on the tired light theory gets rid of the problems that are related to Big Bang, that is, the super velocity problem, the horizon effect, and the problem of the beginning of the Cosmos. Moreover, the model explains the cosmic microwave background radiation as a natural result of the tired light effect, and therefore, Olbers' paradox is disappeared. Based on the tired light theory and together from the cosmological principle, the Cosmos is infinite and eternal.

Acknowledgements

The authors are grateful to Mr. S. Crothers for reading through the paper and making linguistic corrections and modifications and also thankful to his thoughtful suggestions on the chapter. This work was supported by National Basic Research Program of China grants 973 Programs 2015CB857100

Author details

Ming-Hui Shao*, Na Wang and Zhi-Fu Gao
Xinjiang Astronomical Observatory CAS, Urumqi, China

*Address all correspondence to: shaomh@xao.ac.cn

References

[1] Zwicky F. On the red shift of spectral lines through interstellar space. Proceedings of the National Academy of Sciences of the United States of America. 1929;**15**:773-779

[2] Shao M. The energy loss of photons and cosmological redshift. Physics Essays. 2013;**26**(2):183-190

[3] Hubble E. A relation between distance and radial velocity among extra-galactic nebulae. Proceedings of the National Academy of Sciences of the United States of America. 1929;**15**:168-173

[4] Reber G. Endless Boundless Stable Universe. Hobart: University of Tasmania; 1977

[5] Pecker J, Vigier J. A possible tired light mechanism. In: Hewitt A, Burbidge G, Fang L, editors. Observational Cosmology. Beijing: IAU; 1987. pp. 507-511

[6] Marmet P. A new non-Doppler redshift. Physics Essays. 1988;**1**(1):24-32

[7] Ghosh A. Velocity-dependent inertial induction: A possible tired-light mechanism. Apeiron. 1991;**9**(10):95-119

[8] Assis A. On Hubble's law of redshift, Olbers' paradox and the cosmic background radiation. Apeiron. 1992;**12**: 13-29

[9] Rabounski D. An explanation of Hubble redshift due to the global non-holonomity of space. Progress in Physics. 2009;**1**:L1-L2

[10] Sorrell W. Misconceptions about the Hubble recession law. Astrophysics and Space Science. 2009;**323**:205-211

[11] Mamas D. An explanation for the cosmological redshift. Physics Essays. 2010;**23**(2):326-329

[12] Wilson O. Proportionality of nebular red shifts to wave length. Publications of the Astronomical Society of the Pacific. 1949;**61**:132-133

[13] Espey B, Carswell R, Bailey J, Smith M, Ward M. H_α emission lines in high-redshift quasars. The Astrophysical Journal. 1989;**342**:666-676

[14] Schmidt M, Matthews T. Redshifts of the quasi-stellar radio sources 3C47 and 3C147. The Astrophysical Journal. 1964;**139**:781-785

[15] Nishihara E, Yamashita T, Yoshida M, Watanabe E, Okumura S, Mori A, et al. Redshift differences between the balmer and [OIII] λ5007 lines in high-redshift quasars. The Astrophysical Journal. 1997;**488**:L27-L30

[16] Assis A. A steady-state cosmology. In: Arp H, Keys C, Rudncki K, editors. Progress in New Cosmologies: Beyond the Big Bang. New York: Plenum Press; 1993. pp. 153-167

[17] Adam M. Interferometric measurements of solar wavelengths and an investigation of the Einstein gravitational displacement. Monthly Notices of the Royal Astronomical Society. 1948;**108**:446-464

[18] LoPresto J, Schrader C, Pierce A. Solar gravitational redshift from the infrared oxygen triplet. The Astrophysical Journal. 1991;**376**:757-760

[19] Chastel A, Heyvaerts J. Perturbations of Pioneer 6 telemetry signal during solar occultation. Nature. 1974;**249**:21-22

[20] Merat P, Pecker J, Vigier J. Possible interpretation of an anomalous redshift observed on the 2292 MHz line emitted by Pioneer-6 in the close vicinity of the

solar limb. Astronomy and Astrophysics. 1974;**30**:167-174

[21] Hubble E. The Observational Approach to Cosmology. Oxford: Oxford University Press; 1937

Chapter 3

Model of an Evolving and Dynamic Universe: Creation without a Big Bang

Dietmar E. Rothe

Abstract

This chapter describes a non-relativistic, dynamic universe that *evolves* through continuous transformations of energy forms in contradistinction to a *devolving* Big Bang model. The new cosmology is accurately consistent with Hubble's galactic redshifts, interpreted as simple Doppler shifts of fleeing galaxies, and as viewed from any arbitrary observer in the universe. It postulates the existence of repulsive electromagnetic (EM) force fields between galaxies, while maintaining the purely gravitational dynamics within each galaxy. Observed cosmic redshifts of galaxies and their apparent velocities and accelerations are matched, if galactic cores are assumed to have an unbalanced electric charge of 3×10^{32} C for an average galaxy with a mass of 4×10^{41} kg. Valid arguments are presented for the probable existence of intergalactic EM fields emanating from Supermassive Black Holes (SMBHs) in galactic cores. Special sections in this chapter are devoted: (a) to suggest plausible sources of new matter creation, (b) to discuss how Quasi-stellar Objects (QSOs or Quasars) can fit into this cosmological model, and (c) to counter critiques of the model.

Keywords: cosmology, expanding universe, gravitational fields, intergalactic EM fields, supermassive black holes, galactic redshift, continuous cosmic renewal, age of universe, Hubble's Law, quasars, QSOs, galactic dynamics, cosmic dynamics, globular clusters, baby galaxies, active galactic nuclei, AGN, Kerr-Newman black holes

1. Introduction

Making use of the high resolving power of the 100-inch telescope at the Mt. Wilson Observatory, Edwin Hubble discovered in 1929 that many so-called nebulae were not relatively close accumulations of gas and dust, but were in fact huge systems of billions[1] of stars located at distances far outside of our own Milky Way star system. He also found that light from these *Galaxies* was redshifted in a linear relationship with their distance. This led to two distinct cosmological theories describing an expanding universe populated by billions of galaxies: First, a *Big Bang* model as proposed by Georges Lemaître and Willem de Sitter in 1930/1931, a universe that explosively expanded from a *singularity* of infinitely high energy

[1]Throughout this text, the word *billion* is considered to be equal to 10^9.

IntechOpen

density, and which continues to unfurl on its own without further energy input; Second, a *Quasi-Steady State* theory that relies on new hydrogen being continuously created in intergalactic space, as proposed in 1948 by Fred Hoyle, a universe that could have existed forever. The two theories were accepted as being plausible alternative cosmologies for over three decades. In 1964, radio astronomers Arno Penzias and Robert Woodrow Wilson succeeded in measuring an isotropic microwave background temperature of intergalactic space as being 3.5 K [1]. The new measurements were interpreted as evidence for the afterglow of a hot early universe, in support of the Big Bang theory and as evidence against any rival steady-state theory. Yet, other explanations for the presence of this cosmic microwave background do exist, as discussed in Section 9. Since the late 1960s, the Big Bang theory has been considered the only viable cosmology. In more recent times, however, more deficiencies in the Big Bang scenario have come to light, and it seems prudent to reconsider alternative models of the universe. The author strongly believes that Quasi-Steady models were dismissed in error and need to be revisited.

The present model describes a dynamic, evolutional universe, in which matter is continually created and dissolved back into non-material energy forms in an ongoing cyclic process. It may have existed forever, or it could have evolved in a gradual process over eons of time. Hubble's [2] galactic redshift is interpreted as a recession of galaxies, propelled by an EM force field acting between galaxies. Stars within galaxies are not affected. In this model, new matter is continuously created in intergalactic space and/or is recycled from matter previously swallowed up by SMBHs, so as to keep the population density of galaxies approximately constant as the universe ages [3].[2] Unlike the process proposed by Hoyle [4], it does not depend on intergalactic gas pressure to drive galaxies apart. The present model may be considered analogous to a Friedmann [5], de Sitter [6] universe with a non-zero cosmological constant Λ. Here we interpret the observed galactic redshifts as simple Doppler shifts, and Hubble's constant is taken as H = 74 km/s per megaparsec.

Stellar evolution within galaxies is quite well understood. Stars continually condense from gases and dust, known to be present in the galactic disc. This matter consists of primordial hydrogen, existing since the genesis of a galaxy, and also of matter more recently ejected cataclysmically from nova and supernova explosions. Life cycles of most stars last from 1 to 20 billion years until their nuclear fuel is exhausted. During their relatively short life spans they slowly spiral toward the galactic center, where they are absorbed into a monster SMBH. It is projected here that star systems evolve into brilliant galaxies, containing typically 10^{11} stars, and then slowly blink out after devolving into burned out cinders that may be referred to as naked SMBHs. The process of accumulating all or most of the galactic mass into a central SMBH may take as much as 100 billion years, or longer.

QSOs may be nascent new baby galaxies forming around *condensed* black-hole matter ejected from SMBHs when these become unstable.[3] They may also be burned-out cinders of extinct galaxies. We thus expect to find quasars at all distances in the cosmos, not just at *cosmological* distances. Note that spectral emission lines from most quasars clearly show evidence of C, O, N, S, Al, Si, and Fe. These should not have been present during initial periods in a Big Bang scenario. Many QSOs appear to be clearly associated with mature galaxies that are less than a billion light years away. Halton Arp [7] has made a lifelong study of QSOs, and he ascribes their large spectral redshift to an *intrinsic redshift* in addition to their Doppler shift.

---

[2]Core elements of this model have previously been derived and presented in Ref. 3. The present version contains additional arguments and evidence in favor of the theory.

[3]See discussion in Section 11.

Here we postulate that SMBHs at galactic centers display a diminutive electric charge, accounting for intergalactic EM fields that drive galaxies apart. Because stars within galaxies are electrically neutral, they remain gravitationally bound within their galaxy in the traditional manner. The present theory is offered with similar sentiments as those expressed by plasma physicist Hannes Alfvén: [8] "*Instead of searching for new laws of physics, we should be trying to find out how to use the ones we already know*."

2. Cosmic dynamics

In the following analysis, we will avoid relativistic formulae, so as to circumvent unnecessary complexity. The author believes that a valid description of the cosmos requires an *absolute* point of view, not one that is limited by the speed of light c for transmission of information.[4] Consider a universe containing a uniform, quasi-steady population of n galaxies per unit volume. As the space between fleeing galaxies increases, new galaxies form to maintain a comparable galactic number density n. We select an arbitrary center C to serve as reference point. Consider now a typical galaxy of mass M and core charge q, lying on the surface of an arbitrary spherical boundary at a distance r from center C (**Figure 1**).

For analyzing the motion of this galaxy, we label the volume enclosed by the spherical boundary as Region I, everything outside as Region II. Because of symmetry and because gravitational and electric forces fall off inversely with the square of distance, the cumulative electrical repulsive force F_E on the sample galaxy from all galaxies in Region I and the gravitational attraction force F_G to all galaxies in Region I are the same as if all those galaxies were concentrated at center C. Cumulative gravitational and electric forces on the sample galaxy from Region II outside of the reference sphere cancel out to zero because of symmetry. The electric and gravitational forces are then given by:

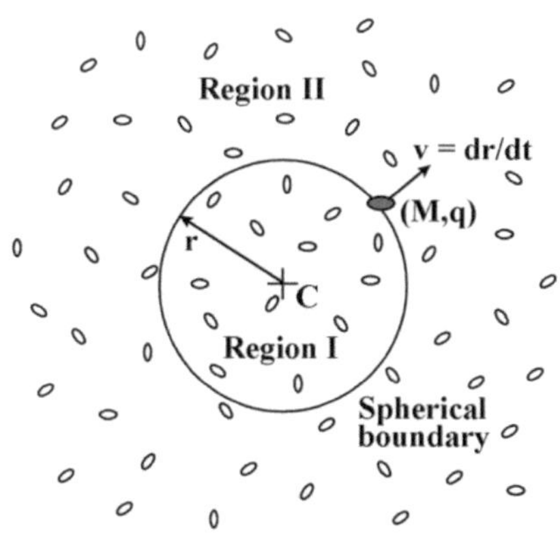

Figure 1.
Division of universe into Region I and Region II to facilitate analysis. This figure has been previously published in: Ref. [3] (http://www.physicsessays.org/browse-journal-2/product/1532-26-dietmar-e-rothe-the-case-for-a-gentler-bang-a-cosmology-of-gradual-creation.html). It is reprinted with permission of Physics Essays Publication.

[4]For spiritually inclined readers, *God* is omnipresent in all that exists. Hence the entire universe remains in his/her/its consciousness at every moment.

$$F_E = \frac{q}{4\pi\varepsilon_0 r^2}\left(\frac{4}{3}\pi r^3 nq\right) = \frac{nq^2 r}{3\varepsilon_0}, \tag{1}$$

and

$$F_G = -\frac{GM}{r^2}\left(\frac{4}{3}\pi r^3 nM\right) = -\frac{4\pi nGM^2 r}{3}, \tag{2}$$

where ε_0 is the electric permittivity of space and G is the universal gravitational constant. Assuming the electric repulsion dominates over the gravitational attraction, the sample galaxy is subjected to a net force F and acceleration directed away from the reference center.

The net force per unit mass on the sample galaxy is then

$$\frac{F}{M} = \frac{n}{3}\left(\frac{q^2}{\varepsilon_0 M} - 4\pi GM\right)r. \tag{3}$$

At time t, the net repulsive force is assumed to have imparted a velocity v = dr/dt to the sample galaxy. From Newton's second law of motion and Eq. (3), we write

$$\frac{F}{M} = \frac{dv}{dt} = v\frac{dv}{dr} = H^2 r, \tag{4}$$

where

$$H = \sqrt{\frac{n}{3}\left(\frac{q^2}{\varepsilon_0 M} - 4\pi GM\right)}. \tag{5}$$

Solving differential Eq. (4) and applying the boundary condition that $v = 0$ when we shrink the arbitrary reference sphere to zero, we find

$$v = Hr. \tag{6}$$

This is Hubble's Law, consistent with observation. The parameters contributing to Hubble's constant are given by Eq. (5). Solving this equation for the galactic charge required to account for the observed cosmic expansion, we find

$$q = \pm\sqrt{M\varepsilon_0\left(\frac{3H^2}{n} + 4\pi GM\right)}. \tag{7}$$

Note that both, velocity and acceleration of galaxies are proportional to r. This naturally indicates that the flight of galaxies has always been accelerating. Evidence to that effect was first reported in 1998 by Schwarzschild [9], giving rise to the need for *dark energy* in the Big Bang model.

Assuming 2.4×10^{11} galaxies exist within a radius of 14.25 Gpc, we derive an average galactic number density of n = 6.7×10^{-70} galaxies per m^3. Taking the average mass of a galaxy as M = 4×10^{41} kg and Hubble's constant as H = $2.4 \times 10^{-18}\ s^{-1}$, we obtain $q = \pm 3 \times 10^{32}$ coulombs per galaxy, which works out to one elementary charge, i.e. one extra proton or electron for every 1.3×10^{17} atomic mass units (nucleons) in the galaxy. We only need such a slight deviation from neutrality because electric repulsion between protons is over 10^{36} times stronger than gravitational attraction between nucleons. The minimum electric charge necessary to balance gravitation is approximately one elementary charge for every 10^{18} nucleons.

It is generally taken for granted that the number of positively charged subatomic particles in the universe is exactly balanced by an equal number of negatively charged particles, although this assumption may be challenged. So, we need to explain how SMBHs can become electrically charged. First, it is possible that charge may not be conserved in the extremely compressed state of matter inside a black hole. Second, charges may become irretrievably separated during the accretion process onto a rapidly spinning SMBH in the presence of strong magnetic fields. We only need a minute imbalance in charge, and hence only a minute preference for particles of different charge and mass to be captured. If it is slightly more probable for protons to enter the SMBH horizon than for electrons, then a net positive charge would accumulate inside the event horizon and a net negative charge outside.

In a paper by Price and Thorne [10], they verified analytically that a black hole's event horizon can be considered an electrically conducting membrane with a resistivity of 377 ohm, the dynamic impedance of space, $Z_0 = \sqrt{\mu_0/\varepsilon_0}$, where μ_0 is the magnetic permeability of *empty* space. Such a conductive event horizon can easily shield the internal charge from outside view by induced electric currents in the conducting event horizon, so that externally the SMBH appears to be negatively charged.

3. Continuous renewal process

In the above derivations the number density n of galaxies in space has been assumed to be independent of time. In an *expanding* universe, this condition requires new matter to enter physical space between galaxies on a continuing basis. Note that the present model does not require space itself to expand, even though that idea, if true, could easily be incorporated into the model. New matter then condenses into new galaxies. In our model, we can estimate what the influx of new, or recycled, matter energy per unit volume of space would need to be. Consider Region I in **Figure 1** to contain N galaxies within the spherical volume V. The rate of expansion of this hypothetical spherical volume is then given by

$$\frac{dV}{dt} = 4\pi r^2 \frac{dr}{dt} = 4\pi H r^3. \tag{8}$$

The number of new galaxies needed per unit volume per unit time is:

$$\left(\frac{1}{V}\right)\frac{dN}{dt} = \left(\frac{n}{V}\right)\frac{dV}{dt} = 3nH = 4.82 \times 10^{-87} \text{ galaxies m}^{-3}\text{s}^{-1}. \tag{9}$$

This corresponds to only one new proton per cubic meter every 30 billion years, or one solar-mass star per mega-parsec cubed every year. Continuous creation of new matter takes place most slowly. Similarly, the rate of acceleration in the cosmic expansion, driven by new matter energy, is only H^2r as given by Eq. (4), where $H^2 = 1.77 \times 10^{-16}$ km/s^2 per mega-parsec.

Formation of new matter may occur in different ways:

1. Subatomic particles may arise out of the stressed quantum vacuum field. Protons uniting with electrons form hydrogen atoms, the basic stuff in the cosmos.

2. New matter energy bleeding into this universe from *parallel universes* has also been seriously considered by cosmologists. But that speculation is not needed for the model discussed here.

3. Alternatively, matter could be recycled by aging SMBHs, known to exist in the cores of mature galaxies. Based on observational evidence, this appears to be the most probable source.

High-velocity jets of particles, and apparently also of QSOs, shooting out of magnetic poles of active galactic nuclei (AGN) are common in the universe. These phenomena strongly imply the presence of electromagnetic fields emanating from galactic cores. When spinning SMBHs are assumed to inhabit galactic centers, most astronomers would consider that they play a central role in the evolution of QSOs, of AGNs, and of galaxies. Certain globular star clusters also seem to be possible steppingstones in the formation of *Baby Galaxies*. Entire globular star clusters, ejected at high speed from active galactic nuclei, have recently been observed. Black holes have been found in many global star clusters, and stars seem to be orbiting in an organized way around the center of star clusters. More details of these new observations are reviewed in following Sections 4 and 5.

4. QSOs, globular clusters, and baby galaxies

Relatively recent press releases by NASA/JPL-Caltech reported new observations obtained via the *Galaxy Evolution Explorer* (GALEX) satellite. The new findings took astrophysicists and cosmologists by surprise [11]. Equipped with highly sensitive UV detectors, the GALEX spacecraft was launched in 2003 to scan the heavens in search of young galaxies in their formation stage at the edge of the observable universe, more than 10 billion light years (ly) from us. Newly forming, young galaxies are about ten times brighter at ultraviolet wavelengths than mature galaxies. The unexpected discovery was that such baby galaxies have not only existed around 10 billion years ago, as per accepted cosmology, but dozens of them were found to exist relatively close to us at inferred distances ranging from 2 to 4 billion ly away, with some possibly being as young as 100 million years. The impact of this discovery to cosmologists should be huge. They may need to rethink their cosmological perspectives. As nature is able to renew itself indefinitely, the universe may also be evolving by renewing itself in an ongoing process, instead of devolving toward an ultimate entropy death.

Other new observations imply that globular star clusters and dwarf spheroidal galaxies are in an evolutionary sense related to galaxies. They may well be precursory star formations that gradually grow into those newly discovered baby galaxies mentioned above. Globular star clusters and dwarf galaxies may be found around most mature galaxies, including our Milky-Way galaxy. Spheroidal galaxies also appear to contain great amounts of dark matter, an indication of the presence of massive black holes there [12].

A similar discovery, which may revolutionize our understanding of the nature and origin of globular star clusters, shocked astrophysicists in 2014, *upsetting 40 years of theory* about these spherical star formations [13]. A team led by Prof. Tom Maccarone of Texas Tech University in collaboration with Maximilian Fabricius at the Max Planck Institute for Extraterrestrial Physics in Garching, Germany, *recently found a surprise* when new observations ascertained that stars at centers of older star clusters *rotated around a common axis instead in random orbits, as once thought*. They also found that many stars in these clusters consisted of relatively young stars, and not just of old stars as previously perceived. The Texas Tech team had been investigating globular star clusters for many years in search of intermediate size black holes. In 2007, they made the first discovery of a stellar mass black hole in a star cluster near a neighboring galaxy NGC4472. In 2012 astronomers from several universities, using the *Very Large Radio Telescope Array* found a binary black hole in the core of the M62 star cluster [14]. Since then many more black holes, containing many solar masses, have been detected in numerous extragalactic globular clusters. We may conclude that certain globular clusters can evolve into baby galaxies. We should also note that in 2014 astronomers discovered a globular star cluster of thousands of stars, near the supergiant elliptical galaxy M87 in the Virgo Cluster. This globular star cluster appears to have been ejected in its entirety from the supermassive core of M87 at a velocity of over 2000 km/s. [15] Even though M87 weighs as much as 6 trillion (10^{12}) Suns, the globular star cluster will escape the gravitational pull of its source galaxy. However, there may be another more direct alternative source for baby galaxies and QSOs, as described below.

5. QSOs born out of mature galactic cores

The first bright quasi-stellar objects were observed and recorded around 1960. All of them emit strong EM radiation from radio waves to X-rays and Gamma rays [16, 17], and all of them exhibit a high redshift in their emission spectra. If interpreted as Doppler shifts or as cosmological expansion redshifts, they appear to be receding from us at extreme speeds. As per contemporary models of the universe, this would put them far out into the cosmos, where cosmologists believe to see them as early precursors in the evolution of the first galaxies some 10 billion years ago. Cosmologists did not expect to find any QSOs nearby.

Present understanding is that QSOs are powered by SMBHs. But not until the Hubble telescope in the 1980s could detect faint traces of matter around QSOs extending approximately 10 ly out from the bright quasi-stellar objects, suggestive of galactic spiral arms, were QSOs considered *Active Galactic Nuclei* (AGNs) in young galaxies in formation.

In general,[5] AGNs are considered to consist of SMBHs containing more than a billion solar masses, surrounded by active accretion discs, where matter spirals into the black hole at relativistic speeds. It should be noted here that gravitational force gradients at the *Event Horizon* of black holes get weaker the more massive and large the SMBH is. When a black-hole mass exceeds about 200 million solar masses, stars sucked into the hole no longer get torn apart near the event horizon. They get swallowed whole. Thus, it is not clear how the accretion discs of AGNs can produce strong X-ray radiation.

Average sizes of active regions of QSOs that are brilliantly bright may typically have estimated radii of half a light year. The Schwarzschild radius ($r_S = 2GM/c^2$) of a SMBH's *event horizon*, containing 10^9 solar masses is only $r_S = 3.12 \times 10^{-4}$ light years. Hence, the visible radiation emitted from the accretion disc appears to extend to at least a thousand r_S, or to 0.31 ly from the center. Applying Newtonian/Keplerian mechanics, we find that matter orbiting an SMBH at that distance and emitting EM radiation (at $r_E = 10^3\ r_S$) has an orbital speed of $v = (GM/r_E)^{1/2} = 6710$ km/s. However, the radial velocity gradient there is only $dv/dr = 5.7 \times 10^{-13}$ km/s per meter, far too small for any thermal radiation in the visible or X-ray spectrum to be

[5]AGNs were originally believed to exist only in large mature galaxies.

produced. If there were any light emitted from the accretion disc at $r_E/r_S = 10^3$, any gravitational redshift[6] would only be $z_G = 5 \times 10^{-4}$, and it would not significantly contribute to the observed redshifts of QSOs. For gravitational redshifts to play a significant role, the radiation must then come from locations r_E closer to the event horizon than $r_E = 10r_S$ where z_G is greater than 0.054. The source for the brilliance of QSOs, which is assumed to be greater than the combined radiation from 10^{14} stars, remains an unanswered enigma, if QSOs are at cosmological distances of 4 to 10 billion light years from us.

Astrophysicists are generally at a loss to explain what the power source is that drives the immense radiative power output from a quasar. It would have to be much more efficient in converting matter energy and kinetic energy to radiation energy than nuclear fusion. One concept discussed by astrophysicists consists of a complete conversion of gravitational potential energy of matter to radiation at the edge of the SMBH. Yes, gravitational potential energy of matter falling toward a black hole is converted to kinetic energy, while also gaining relativistic mass. But what are the conditions and processes necessary to convert this energy into electromagnetic radiation? Any required non-thermal condition is not likely to be met outside of the event horizon of a SMBH. The needed compression of in-falling matter would only occur well within the black hole event horizon, wherefrom radiation cannot escape, unless perhaps the SMBH is a fast spinning, electrically charged *Kerr-Newman* black hole [18]. The enigma can however be significantly alleviated, when we consider QSOs to be much closer to home; i.e. if at least the very bright ones are a 100 times closer. If they were 100 million ly away, instead of 10 billion ly, the apparent visible area of their active nuclei would be 10,000 times smaller, and their much smaller radiative power output more explicable with conventional physics. The radius of the active region, instead of being half a light year, would be 5×10^{-3} ly and would only be approximately 10 times larger than the event horizon of the central SMBH. In that case, a matching accretion disc, mostly spinning at hyper-velocities, can easily supply the lesser radiative power output, if QSOs are much closer than previously believed. Making that assumption would vindicate astronomer Arp's hypothetical claims based on decades of meticulous and accurate observations and data. His deductions [19] include:

1. High redshift QSOs are often closely associated with lower redshift mature galaxies with active nuclei and are, therefore, at similar distances.[7]

2. QSOs are ejected at hyper-velocities from AGNs of mature galaxies.

3. High redshifts of QSOs have two component parts: the major part is based on an *Intrinsic Redshift*, in combination with a lesser Doppler shift.

4. QSOs ejected from a parent galaxy are lined up with their intrinsic redshifts decreasing with distance from their source galaxy.

5. QSOs evolve into normal galaxies over eons of time.

---

[6]See Eq. (14) below.

[7]For example, a high-redshift quasar is clearly in front of the low-redshift galaxy NGC 7319. Many others are connected by luminous bridges to a parent galaxy.

6. Alignment and motion of QSOs

In 1969, Morley Bell of NRC, Canada published a statistical analysis [20] of 150 QSO redshifts, which suggests that *there are groupings of QSOs extending over large areas of sky*, with the inference that *QSOs are not at cosmological distances.* This paper is significant, because it was written 21 years before the *Hubble Telescope* was launched into orbit and before newly discovered quasars were relegated to the far corners of the universe. Hence, the 150 QSOs listed at that time by Burbidge and Burbidge [21] were a select group of quasars with high apparent brightness and relatively low redshifts, indicative of their non-cosmological distances, before skies became cluttered with thousands of more recently discovered QSOs that may truly be located much farther away. Bell presented in his paper statistically significant evidence that these QSOs can be grouped according to their redshift; the groupings being evident from redshift histograms, which indicate population peaks at regular redshift intervals of 0.172 or multiples thereof. He thus sorted the redshifts into twelve groups and found within each group a linear relationship between the redshifts and their angular distance from a somewhat arbitrary group center in the sky. His most far-reaching observation, however, has been the connecting of QSOs into orderly spiral patterns, covering large areas of the sky and having redshifts either increasing or decreasing monotonically along the spirals.[8]

Bell identified each spiral group by two numbers; the first being the number of the peak in his redshift histogram, the second specifying whether the group is in the Northern Hemisphere (near 12 h right ascension) or in the Southern sky (near 0 h right ascension). The same numbering is retained here to afford easy reference to Bell's paper. From the twelve spiral groups identified by Bell, the present author has selected two such spirals, N = (2, 12 h) and N = (1a, 0 h). When considered as 3D helices instead of 2D spirals, these two are the only pair that can reasonably be connected across the galactic equator gap, where any intergalactic QSOs are obscured by the *Milky Way* galactic disc. **Figure 2** shows a plot of these two QSO spirals in a sky chart using Earth equatorial coordinates. It is essentially a copy of Bell's sky chart [Ref. [20], Fig. 9], with spiral groups other than N = (2, 12 h) omitted and with group N = (1a, 0 h) added in the Southern sky.

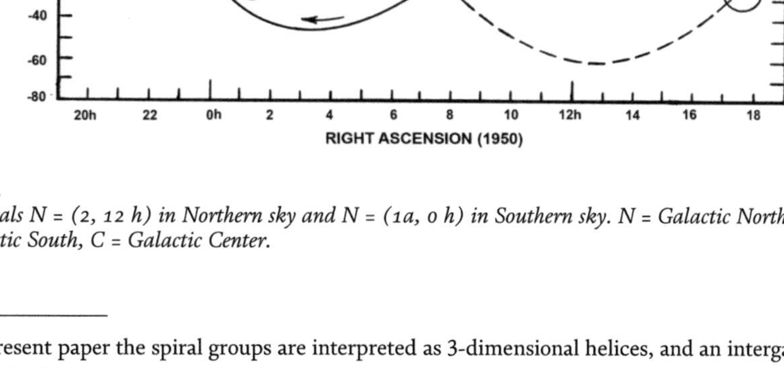

Figure 2.
QSO Spirals N = (2, 12 h) in Northern sky and N = (1a, 0 h) in Southern sky. N = Galactic North, S = Galactic South, C = Galactic Center.

[8]In the present paper the spiral groups are interpreted as 3-dimensional helices, and an intergalactic magnetic field is postulated.

In the diagram, individual QSOs are represented by small circles together with their redshifts. Note how the redshifts decrease in a counter-clockwise direction going outward from the center of the spiral in the Northern sky and then decrease in a clockwise direction toward the center of the Southern spiral. When viewed as a 3D helix, the combinations of the two spirals represent a continuous helical path spiraling in the same direction from north to south. The spiral centers are considered vanishing points as seen from an observer on Earth. The redshifts decrease in the same direction from north to south in both hemispheres. The Northern vanishing point lies in the direction of the Virgo cluster of galaxies, and is very close to the giant elliptical galaxy M85, which is known to have an active nucleus. Thus, the diagram would be consistent with the proposition that the QSOs traveling along their helical path are ejected from an AGN source, as portrayed in **Figure 3**.

The relatively large number of QSOs in the Northern sky spiral belong to a well-defined redshift group, and that spiral curve is unambiguous. Thus, there is a reasonable statistical support in favor of a causal connection between these QSOs. From **Figure 2** it is clear that the QSOs could not have been ejected from the center of our local Milky Way Galaxy. Assuming that the helical path conjecture is correct, this helical path must have intergalactic dimensions, as shown in **Figure 3**. Our entire Milky Way galaxy then lies within the helix, though not on its central axis.

Gravitational and other central fields cannot put moving objects on a helical trajectory. Alternatively, helical trajectories are a common occurrence in magneto-plasma physics, where ions and electrons are found to spiral around magnetic field lines along helical paths. We know that strong magnetic fields are created by AGNs that enable and guide material to be ejected as hyper-velocity jets from their magnetic poles. Sources of these intergalactic magnetic fields can easily be identified and understood, if rapidly rotating SMBHs are present in galactic cores and QSOs, and if their *event horizons* display an electric charge.[9] Thus, electrically charged QSOs fit well into the author's theory presented here. To complete the picture shown in **Figure 3**, the QSOs shown have probably been ejected from a source galaxy (AGN) that is also the source of the magnetic field; i.e. the major velocity vector of the QSOs must be aligned with the magnetic field lines, with a minor transverse velocity component to produce the helical trajectory.

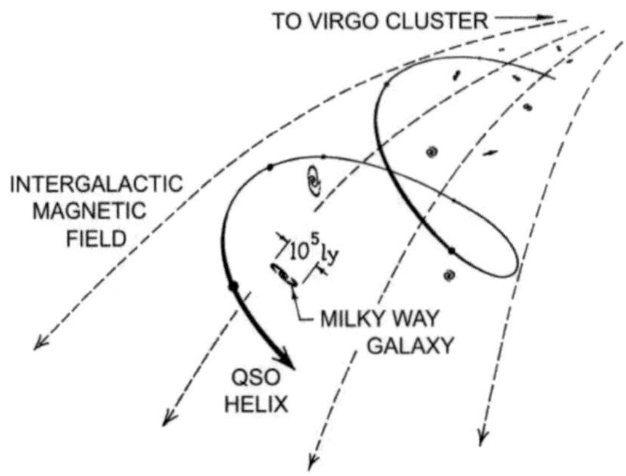

Figure 3.
Helix of electrically charged QSOs within an intergalactic magnetic field originating in the Virgo Cluster of galaxies.

[9]See Section 2: *Cosmic Dynamics*.

According to Hubble's Law, distances of QSOs in the Northern sky in **Figure 2**, as derived from their redshifts, are approximately 3 billion ly from us, whereas in the Southern sky QSOs would be twice as far. Thus, the two groups of quasars could not possibly be connected in any way by their Doppler shifts alone. However, if we accept Arp's suggestion that redshifts of quasars are partly intrinsic and partly Doppler shifts, we can make a case for the two QSO groups being indeed related. From Earth's viewpoint the northern QSOs are approaching us and the ones in the Southern Hemisphere are receding from us. Let the redshift of the approaching ones be z_A and that of the receding ones be z_R. Then we can write the following relations:

$$z_A = z_I - z_V \simeq 0.24 \tag{10}$$

and

$$z_R = z_I + z_V \simeq 0.46, \tag{11}$$

where z_I is the intrinsic redshift and z_V is the redshift due to velocity. By adding and subtracting Eqs. 10 and 11 we get:

$$z_I = \frac{z_R + z_A}{2} \simeq 0.35 \tag{12}$$

and

$$z_V = \frac{z_R - z_A}{2} \simeq 0.12. \tag{13}$$

To an approximate degree this would mean that the speed v of the QSOs along their helical trajectory is around 36,000 km/s, and that 74% of the observed redshift is due to an intrinsic redshift at their closest approach to our galaxy. These numbers are quite consistent with claims made in the author's theory and also with Arp's conjectures. The intrinsic redshifts are consistent with gravitational redshifts for radiation coming from areas close to SMBHs. A back-of-the-envelope analysis to this is given hereunder:

What are typical distances of the QSOs studied?

If the source of QSOs is the AGN galaxy M85, their distances must be less than 60 million ly, the distance to M85. The large geometrical patterns in the sky by the helical trajectory suggest that at closest approach to our galaxy the radius of the helix may be of the order of a million ly. An average geometric mean of the QSO distances may be 8 million ly. By comparison with the generally accepted distance of 3 billion ly, based on a redshift of 0.24, these QSOs may on the average be 400 times closer.

What is the actual size of the active region?

Statistical size estimates of active nuclei in quasars, when at cosmological distances, seem to be at approximately 1 ly in diameter. The corresponding radius of the active region at the closer distance would be 1.2×10^{-3} ly.

What is the expected intrinsic gravitational redshift?

The redshift of light escaping from close to the event horizon of a Schwarzschild black hole is,

$$z_G = \frac{1}{\sqrt{1 - r_S/r_E}} - 1, \tag{14}$$

where r_S is the radius of the black hole and r_E is the radius of the source of the escaping radiation. To explain a gravitational redshift of $z_G = z_I = 0.35$ as calculated in Eq. 12 for a typical QSO, we need r_S/r_E to be 0.45, or $r_E/r_S = 2.22$. The Schwarzschild radius of a 1 billion solar mass black hole is 3.1×10^{-4} ly. A fast spinning, charged Kerr-Newman hole of that mass may have an equatorial radius of $r_S = 6 \times 10^{-4}$ ly. Observed radius of the active region of typical QSO at the closer distance is $r_E = 12 \times 10^{-4}$ (see above). We then have $r_E/r_S = 2.0$ as required to show that the intrinsic redshift of 0.35 can easily be explained as a gravitational redshift.

7. Minimum age of the universe

The theory discussed here does not explain how the present structure of the universe came into being. It may have been in existence forever, or it started from a small mini-universe some 200 billion years ago. Arguments that have been presented here essentially describe a dynamic universe that has perhaps existed for an unimaginably longer time than the estimated age of the Big Bang universe. If the universe evolved from a simple beginning, we first need to define how large an initial volume was needed to define a process of expansion that could keep evolving in a systematic manner as described in Section 2. Similarly, we would need to know how large the universe really is now. Defining the radius R_O of the *Observable Universe* by the edge at which the fleeing galaxies reach the speed of light, the extent of this universe, as we observe it, is given by $R_O = c/H = 13.2 \times 10^9$ ly. But we see these farthest observable galaxies where they were 13.2 billion years ago. By the time light from galaxies at the observable edge of the cosmos reaches us, they are now approximately $R_K = 40 \times 10^9$ ly from us. Let R_K be the radius of the *Known Universe*. For obtaining a minimum age of the universe, according to the present theory, a certain minimum original size is needed for Eqs. 4 and 6 to be meaningful, and for H to be approximately constant at its present value.

In Ref. [3] the author has made an analysis for estimating a time period needed for an initial *mini-universe* to evolve into the present state of our universe. He assumed an initial mini-universe consisting of a spherical volume of 1.7×10^{13} ly^3, having radius $r_I = 13{,}000$ ly at time t_I, and containing 20 globular star clusters with electrically charged central black holes.[10] Such a mini-universe may already be quite old, given the time needed for stars to form and to collapse into black holes. Using formulas from Section 2, it can be shown that

$$t_U - t_I = \frac{1}{H}\ln\left(R_K/r_I\right) = 200 \times 10^9 years, \tag{15}$$

where t_U is the present time.

The age of the universe is thus more than 2×10^{11} years. If the *Actual Universe* extends beyond the *Known Universe*, then it would be much older.

8. Expanding space or fleeing galaxies?

The author's mathematics have assumed that n and H are not functions of time. They implicitly assumed that galaxies move through an existing stationary space (Situation A). If the cosmic expansion is one of space itself that carries galaxies with

[10]Evidence of black holes in globular star clusters has been accumulating since 2007 (see Section 4). Additional evidence shows that certain star clusters can evolve into galaxies.

it (Situation B), most of the above arguments would still stand. But what we mean by the *Observable Universe* and the *Knowable Universe* would be different. In an Einsteinian situation B universe, in which the speed of light is an absolute limit,[11] galaxies nevertheless moving away faster than c, could not be observed, because EM signals could never reach us. Even if they could, their redshifted photons would have lost all their energy, dissipating themselves. In a *Maxwellian* situation B universe, EM waves from faster-than-light moving galaxies would also never reach us. In a *Maxwellian* situation A universe, however, EM waves from faster-than-light galaxies moving through a stationary space would eventually reach us, and their Doppler frequencies would appear to be less than half their emitted frequencies (more than twice their normal wavelengths).

As of this date the farthest galaxy detected has a redshift of z = 11. In a linear system obeying Hubble's Law (v_{GAL} = Hr = zc), this galaxy is receding from us at a speed of 11 times the speed of light. It was at a distance r_0 of 1.45×10^{11} ly from us when at time t_0 it emitted the light we are receiving now. This light took 145 billion years to get to us ($\Delta t = t_{NOW} - t_0$). According to Eq. (15),

$$R_{NOW} = r_0 e^{H\Delta t}. \tag{16}$$

This places the galaxy at R_{NOW} = 8.6×10^{15} ly away from us. In a non-relativistic cosmos, the present theory predicts a much older and larger universe than the Big Bang Theory.

The current version of the Big Bang Theory (BBT), augmented with Einstein's General Relativity Theory, attributes the cosmological redshift to the expansion of space itself, which stretches EM waves passing through it. In this theory, the speed of light is never exceeded by the receding galaxies, even as their redshift z goes to infinity. Hence, according to the BBT, the Observable Universe is limited in size to a radius of R_{OBS} = 13.8 Gly, and the *knowable* edge of the universe is limited to R_K = 46 Gly. We note that a space that warps and stretches cannot be a total void, in contradistinction to basic assumptions made in the Special Theory of Relativity. A vacuum space that can warp is equivalent to the contentious and disparaged *luminiferous aether*, assumed to be necessary by Maxwell for propagation of EM waves. Post-modern Quantum Physics recognizes vacuum space as a zero-point, high-energy field and not as a total void. The last word regarding the validity of any cosmological model has to await future discoveries in our understanding of basic reality; discoveries about the nature of space, time, mass, and electric charge.

9. Answers to critiques of the cosmology presented

Critique 1: Steady State universe models have been proven wrong.

Answer: Firstly, the cosmology presented here is anything but static. The universe is shown to be highly dynamic and evolving. Secondly, it cannot be considered wrong just because it is incompatible with the BBT. It should be noted that *quasi-steady state theories* rest on fewer *a priori* assumptions than the BBT. Moreover, they can match most observations with few, if any, adjustable parameters, whereas the BBT critically relies on a growing number of them. Astrophysicist and cosmologist Tom van Flandern [22] has made a list of *The Top 30 Problems with the Big Bang*, a theory which needs introduction of new concepts and new adjustable parameters to make it more compatible with any new observations. Quasi-steady

[11]By definition a contradiction in *relativistic* theories.

state universe models have been mistakenly abandoned on observational data that can be interpreted in alternative ways. See further discussion below.

Critique 2: The theory cannot explain the cosmic microwave background radiation (CMB).

Answer: The CMB discovered in 1964 corresponds to a uniform isotropic background temperature of space of 2.7 K. It was heralded as clear proof that the BBT is the most realistic cosmology. But as Van Flandern states, *the CMB makes more sense as the limiting temperature of space heated by starlight than as the remnant of a fireball* [22]. Astrophysicist Arthur Eddington [23] had already determined in 1926 that interstellar space would have a limiting temperature of 3.2 K. In 1933, Ernst Regener [24] predicted that intergalactic space has a temperature of 2.8 K. These predictions were made for steady-state models, well in advance of the BBT. Yet they accurately predicted what is observed now as the CMB. Conversely, early pioneers and proponents of the BBT failed to predict the measured CMB temperature correctly. George Gamow predicted a CMB temperature, as the afterglow of the Big Bang plasma, of 7 K in 1955, then updated it to 50 K in 1961. The discrepancy is highly significant, because energy radiated from a *black body* per unit surface area per second is 118,000 times higher at 50 K than it is at 2.7 K. The BBT failed to explain the temperature of the CMB. It also fails to explain other aspects of the observable universe. For example, it cannot explain why space does not seem to expand within galaxies and within galaxy clusters.

COBE satellite data of the CMB also conflicts with BBT expectations in two ways. Firstly, the microwave background varies only by less than one part in a hundred thousand and cannot explain how super-clusters, sheets, walls, and filaments separated by immense voids in the large-scale distribution of galaxies could have formed in less than 100 billion years [25]. Secondly, the COBE temperature data displays unequivocal dipole anisotropy [26], believed to be due to the Earth's motion relative to a *co-moving* reference space that follows the general expansion of the universe. This nonconforming motion of the Earth amounts to 370 km/s. Measurement of the motion of Earth relative to a cosmic background in effect defines a preferred reference frame, the existence of which was prohibited by Einstein.

Critique 3: The theory confuses cosmological redshifts with Doppler shifts.

Answer: The present theory treats observed redshifts as Doppler shifts and as intrinsic redshifts, such as gravitational redshifts (specifically as applied to quasars) [7]. See 3rd and 4th paragraphs in the Introduction. A separate *cosmological redshift* caused by expanding space is not needed.

Critique 4: The mechanism of creation of new matter is not clear.

Answer: True, we do not understand the exact mechanisms by which energy takes on material form.[12] Matter is itself an energy form, and space itself contains enormous amounts of energy. Even the BBT, based on expanding and distortable space, cannot explain how new energy is produced for the creation of new space, as the universe expands. This is a paradox that seems to violate energy conservation laws.

The main text of the present theory hints at different possible *mechanisms of creation of new matter* (see Sections 3 to 7).

Critique 5: Since galaxies have different mass, they would have different electric charges and the universe should be very inhomogeneous.

Answer: The universe is indeed very inhomogeneous at all levels, an observational fact that the BBT cannot explain, but which is easily explained by the present theory.

---

[12] We know it happens, as demonstrated by electron-positron pair production from gamma rays.

Critique 6: In an accelerating universe the number density of galaxies must vary with time.

Answer: In the BBT, the number density of galaxies varies with time, with or without acceleration, except when measured against an expanding *co-moving reference frame*, which does not fit the scientific definition of a reference frame. Within precepts of the present theory, acceleration of galaxies is a natural consequence of electric fields. Continuous new creation of matter is only needed to keep the number density of galaxies approximately constant. The theory is amendable, should new observations prove that H or n change with time. Note also that in the present model the velocity differential between galaxies one million parsecs apart increases by only 6 km/s in a billion years.

Critique 7: The theory does not resolve the *dark energy* problem.

Answer: The theory has no need for the presence of *dark energy*, other than the energy contained in electric fields and in the underlying zero-point quantum vacuum field.

10. Author's suggestions

As this composition exemplifies, detailed knowledge about the inner workings and relationships of SMBHs and quasars is a cosmological key to understanding the universe we live in. We must be free and able to contemplate and pursue fresh ideas and to let go of unworkable old traditional concepts. For example, we will not solve the mysteries of AGNs until we really understand what happens inside the *event horizon* of SMBHs. For making progress along these lines, I offer a few suggestions:

1. Recognize that in the balance of nature, there can be no physical *Singularities* or *Infinities*. The mere concepts are inherently contradictory within themselves and cannot exist in a tangible reality.

2. Give electromagnetic forces and effects proper importance in any meaningful cosmology. Most powerful electromagnetic forces in addition to relatively weak gravitational forces need to be acknowledged.

3. Encourage and engage the brightest plasma physicists to become *Plasma-Cosmologists*.

11. Passing thoughts

There appears to be sufficient evidence to believe that AGNs in massive, mature galaxies are driven by events occurring in central SMBHs; that these black holes have gravitationally accrued matter, primarily in the form of hydrogen atoms, hydrogen ions (protons), and electrons over eons of time; that this matter entered the event horizon via a rapidly spinning (near the speed of light) equatorial accretion disc; that this spinning galactic nucleus causes a strong inter-galactic magnetic field; that this active nucleus periodically ejects material jets from its magnetic poles at relativistic velocities.

We can perhaps get a better picture of structures and processes associated with AGNs by comparison with other similar processes at smaller scales more familiar to us. An axiom of the so-called *Perennial Philosophy* proclaims: *As above, so below*. This is generally understood to mean: *Structures and processes in the Macrocosm simulate those present in the Microcosm*. Certainly, the same basic laws of physics are valid in both domains.

Compare the strong magnetic field effects of an active galactic nucleus (AGN) with the weak magnetic field of planet Earth. Earth's solenoidal magnetic field catches and entraps high-energy ionic particles from the *Solar Wind* to form the *Van Allen Belts*, two somewhat distorted toroidal belts surrounding the globe. The inner belt consists primarily of high-energy protons. The much larger outer belt holds mostly high-energy electrons. These particles are forced into helical pathways encircling magnetic field lines. Electrons and protons coming from the sun or from the Van Allen belts can enter the Earth's ionosphere at the magnetic poles, causing aurora displays. Trapped ions in each belt reach the outer layers of the atmosphere at the poles, because there the guiding magnetic field lines curve inwards to the planet's surface.

We should expect similar processes to occur in the ultra-strong magnetic field produced by an AGN, only immensely more intense. As generally believed, an AGN should have a central SMBH that accumulates matter via an accretion disc, where strong gravitational forces impel entire stars, dust, and other non-ionized matter[13] to swirl into the black hole. Because most galactic matter had been orbiting around the galactic core in one direction already, conservation of angular momentum will speed this matter up to relativistic velocities. For this reason, all SMBHs must be rapidly spinning oblate Kerr-type holes.

Observational evidence seems to indicate that AGNs and QSOs are enveloped in strong magnetic fields, with the magnetic poles acting as gates that enable matter to be ejected as material jets at relativistic speeds. In addition to neutral accretion discs, AGNs should also be surrounded by electron belts and by positive heavy-ion belts closer to the event horizon. If, as suggested in Section 2, protons and alpha particles find their way through the event horizon more easily than electrons, then the AGNs would present a surplus negative charge to the outside world. Electric and magnetic fields from the corresponding surplus positive charge within the SMBH are neutralized by electric currents in the event horizon and cannot go beyond it [10]. Hence, a spinning AGN or QSO will appear to be electrically charged and will be the source of its own EM fields.

To explain how jets of matter and QSOs can be ejected out of AGNs, consider galaxies as the *atoms of the cosmos*. In analogy, charged atomic nuclei have a mass limit beyond which they become unstable and undergo radioactive decay, emitting protons, alpha particles, or ejecting whole portions of nuclear material, as in nuclear fission processes. Similarly, spinning and electrically charged black holes may also have an upper mass limit. They may not be able to compress matter much beyond neutron-star density before they become unstable. Remember that electrical repulsion between protons is 10^{36} times stronger than their gravitational attraction. Whenever, during the mass accumulation process, AGNs reach the point of inner instability they will eject nuclear material and/or burp out QSOs. The periodicity of such events may be responsible for the apparent redshift *quantization* observed with quasars. QSOs appear to be *born* with their own relatively small but charged SMBHs. As these SMBHs become more massive and larger with time, as they travel along their trajectories, the gravitational field gradients at their event horizons become smaller, accounting for their decreasing gravitational redshifts, the farther they travel from their mother AGN.

[13]Like hydrogen atoms and molecules, whose emission lines show up in QSO spectra.

12. Summary

This study describes an expanding cosmos that maintains an approximately uniform concentration of galaxies. It explains many observed mysteries, and it addresses inconsistencies in other theories. Galactic velocities and accelerations increase linearly with distance from any observer. Such a universe is shown to be older than 200 Giga-years. The theory has no need to search for large amounts of *dark matter* to make the universe *flat*, as there is no overriding requirement for it to be so. We do not have to invent unproven conditions and mechanisms, such as near-infinite energy densities and near-infinite accelerations (as in inflationary periods), to explain the initial phases of creation, and we have no irreconcilable conflicts with observational evidence. The above analysis of the proposed theory shows that the evolution of QSOs may be the most probable creation process needed to keep the number density of galaxies in the cosmos approximately constant in time in an *expanding* cosmos.

Author details

Dietmar E. Rothe
Rothe Technical Research, Cardiff-by-the-Sea, California, USA

*Address all correspondence to: dietmarr@earthlink.net

References

[1] Wilson RW, Penzias AA. Isotropy of cosmic background radiation at 4080 megahertz. Science. 1967;**156**(3778): 1100-1101. DOI: 10.1126/science. 156.3778.1100

[2] Hubble E. The luminosity function of nebulae II. The luminosity function as indicated by residuals in velocity-magnitude relations. The Astrophysical Journal. 1936;**84**:270-295. DOI: 10.1086/ 143764

[3] Rothe DE. The case for a gentler bang: A cosmology of gradual creation. Physics Essays. 2016;**29**(4):582-587. DOI: 10.4006/0836-1398-29.4.582

[4] Hoyle F. A new model for the expanding universe: Monthly notices Roy. Astronomical Society. 1948;**108**: 372. DOI: 10.1093/mnras/108.5.372

[5] Friedmann AA. Über die Krümmung des Raumes. Zeitschrift für Physik. 1922;**10**:377-386. DOI: 10.1007/ BF01332580

[6] De Sitter W. The expanding universe. Discussion of Lemaître's solution of the equations of the inertial field. Bulletin of the Astronomical Institutes of the Netherlands. 1930; 5(193):211-218

[7] Arp H. Quasars, Red Shifts and Controversies. Berkeley, CA: Interstellar Media; 1987

[8] Alfvén H. Plasma physics applied to cosmology. Physics Today. 1971;**24**(2): 28-31. DOI: 10.1063/1.3022564

[9] Schwarzschild B. Very distant supernovae suggest that cosmic expansion is speeding up. Physics Today. 1989;**51**(6):17. DOI: 10.1063/ 1.882267

[10] Price RH, Thorne KS. The membrane paradigm of black holes. Scientific American. 1988;**258**(4):69-77

[11] Martin DC, Heckman TM, Rich RM, Thilker D. Various Reports 2003-2013. Available from: http://www.galex. caltech.edu

[12] Van den Bergh S. Globular clusters and dwarf spheroidal galaxies. Monthly Notices of the Royal Astronomical Society: Letters. 2008;**385**(1):1-2. DOI: 10.1111/j.1745-3933.2008.00424.x

[13] Maccarone TJ, Fabricius M. Globular Clusters Rotate at Heart. Available from: http://www.astronomy. com/news/2014/05/globular-clusters-rotate-at-heart

[14] Chomiuk L, Strader J, Maccarone TJ, et al. A radio-selected black hole x-ray binary candidate in the Milky Way globular cluster M62. The Astrophysical Journal. 2013;777:69-78. DOI: 10.1088/0004-637X/777/1/69

[15] Caldwell N. Entire Star Cluster Thrown Out of its Galaxy. Cambridge, MA: Harvard-Smithsonian Center for Astrophysics (CfA); 2014. Available from: https://www.cfa.harvard.edu/ news/2014-09

[16] MAGIC Collaboration, Albert J, et al. Very-high-energy gamma rays from a distant quasar: How transparent is the universe? Science. 2008;**320**:1752-1754. DOI: 10.1126/science.1157087 Available from: http://www.sciencemag.org/ contnet/320/5884.full.html

[17] Magic Collaboration, Ahnen ML, et al. Detection of very high energy gamma-ray emission from the gravitationally-lensed blazar QSO B0218+357 with the MAGIC telescopes. Astronomy & Astrophysics. 2016;**595**:A98

[18] Punsly B. High-energy gamma-ray emission from galactic Kerr-Newman black holes, I. The central engine. The Astrophysical Journal. 1998;**498**(2): 640-659. DOI: 10.1086/305561

[19] Arp H. Quasar creation and evolution into galaxies. Journal of Astrophysics and Astronomy. 1997; **18**(4):393-406. DOI: 10.1007/ BF02709331

[20] Bell MB. Clusters of quasi-stellar objects. Nature. 1969;**224**:229-234

[21] Burbidge GR, Burbidge EM. Red-shifts of quasi-stellar objects and related extragalactic systems. Nature. 1969;**222**: 735-741

[22] Van Flandern T. The top 30 problems with the Big Bang. Infinite Energy. 2002;**46**(8):10-15

[23] Eddington A. Diffuse matter in space (Ch. 13). In: The Internal Constitution of the Stars. New York: Dover Publications; 1959. pp. 371-377 (1st Edition 1926)

[24] Regener E. Der Energiestrom der Ultrastrahlung. Zeitschrift für Physik. 1933;**80**:666-669. DOI: 10.1007/ BF01335703

[25] Lerner EJ. The Big Bang Never Happened: A Startling Refutation of the Dominant Theory of the Origin of the Universe. New York: Vintage Books; 1992. ISBN 9780679740490

[26] Kogut A, Lineweaver C, Smoot GF, et al. Dipole Anisotropy in the COBE DMR first-year sky maps. The Astrophysical Journal. 1993;**419**:1. DOI: 10.1086/173453

Chapter 4

Primordial Magnetic Fields and the CMB

Héctor Javier Hortúa and Leonardo Castañeda

Abstract

The origin of large-scale magnetic fields is one of the most puzzling topics in cosmology and astrophysics. It is assumed that the observed magnetic fields result from the amplification of an initial field produced in the early Universe. If these fields really were present before the recombination era, these could have some effects on big bang nucleosynthesis (BBN) and electroweak baryogenesis process, and it would leave imprints in the temperature and polarization anisotropies of the cosmic microwave background (CMB). In this chapter, we analyze the effects of a background primordial magnetic field (PMF) on the CMB anisotropies and how we can have sight the mechanisms of generation of these fields through these features. We start explaining briefly why primordial magnetic fields are interesting to cosmology, and we discuss some theoretical models that generate primordial magnetic fields. Finally, we will show the statistics used for describing those fields, and by using CLASS and Monte Python codes, we will observe the main features that these fields leave on the CMB anisotropies.

Keywords: primordial magnetic fields, CMB, inflation, early universe

1. Introduction

Magnetic fields are ubiquitous in the Universe. Even if the origin of these fields is under debate, it is assumed that observed fields were originated from cosmological or astrophysical seed fields and then amplified during the structure formation via some astrophysical mechanism [1]. If these fields really were present before the recombination era, these could have some effects on big bang nucleosynthesis (BBN) and electroweak baryogenesis process and leave imprints in the temperature and polarization anisotropies of the cosmic microwave background (CMB) [2]. Since PMFs affect the evolution of cosmological perturbations, these fields might imprint significant signals on the CMB temperature and polarization patterns and produce non-Gaussianities (NG) [3]. As a matter of fact, PMFs introduce scalar, vector, and tensor perturbations that affect the CMB in many ways. For instance, the scalar mode generates magnetosonic waves which influence the acoustic peaks and change the baryon fraction; vector mode contributes notably in scales below the Silk damping, and tensor mode induces gravitational waves that affect large angular scales [4, 5]. Further, helical PMFs produce parity-odd cross correlation which would not arise in the standard cosmological scenario [6, 7]. Recently, enough CMB experiments like Planck and Polarbear have presented new limits on the amplitude of PMFs using temperature and polarization measurements that offer the possibility of investigating the nature of PMFs, and it is expected with future CMB polarization experiments like CMB-S4 and Simons Observatory, among others, to improve significantly the constraints to the helicity of PMFs and NG and to be able to provide a new insight into the early Universe [2].

2. A primordial origin

Cosmological scenarios describe the generation of magnetic fields in the early Universe (so-called primordial magnetic field), approximately prior to or during recombination, i.e., $T > 0.25$ eV. At the same time, cosmological scenarios can be classified into two categories: inflationary and post-inflationary magnetogenesis. The first scenario generates PMFs correlated on very large scales during inflation, although the breaking of conformal invariance of the electromagnetic action is needed in order to obtain the suitable seed field. Besides, these kinds of models also suffer from some problems such as backreaction and the strong coupling [8]. On the other hand, post-inflationary scenarios consider PMFs created after inflation via either cosmological phase transitions or during the recombination era (Harrison's mechanism) [2]. However, these will lead to a correlation scale smaller than the Hubble radius at that epoch; thus a suitable field cannot be generated unless we consider another dynamical effect, for instance, helicity, which under certain conditions produces transference of energy from small to large scales required to explain the observational large-scale magnetic fields [9]. We will briefly summarize some models and properties of the cosmological scenarios.

2.1 Inflamagnetogenesis

As mentioned earlier, inflation provides an interesting scenario for the generation of PMFs with large coherence scales. Let us start with the standard free electromagnetic (EM) action, given by [2, 10]

$$S_{EM} = \frac{-1}{4} \int \sqrt{-g} g^{\mu\alpha} g^{\nu\beta} F_{\mu\nu} F_{\alpha\beta} d^4x, \tag{1}$$

where $F_{\mu\nu} = \nabla_\mu A_\nu - \nabla_\nu A_\mu = \partial_\mu A_\nu - \partial_\nu A_\mu$ is conformally invariant and being A_μ the vector potential. By making a conformal transformation of the metric given by $g^*_{\mu\nu} = \Omega^2 g_{\mu\nu}$, the determinant $\sqrt{-g}$ and the contravariant metric change as

$$\sqrt{-g} \to \sqrt{-g^*} = \Omega^4 \sqrt{-g}, \qquad g^{\mu\nu} \to g^{\mu\nu*} = \Omega^{-2} g^{\mu\nu}, \tag{2}$$

and the factors Ω^2 cancel out the action; thus the action of the free EM is invariant under conformal transformations. Since the FLRW models are conformally flat, i.e., $g^{FLRW}_{\mu\nu} = \Omega^2 \eta_{\mu\nu}$, being $\eta_{\mu\nu}$ the Minkowski metric, one can transform the electromagnetic wave equation into its flat version [2, 11]. Therefore, it is not possible to amplify the EM field fluctuations, and this leads to an adiabatic decay of the EM field as $\sim 1/a^2$ with the expansion of the Universe. Hence, inflamagnetogenesis requires the breaking of conformal invariance of the EM action in order to amplify EM waves from vacuum fluctuations [12–14]. A multitude of possibilities have been considered for this purpose, and some of them are illustrated in the action

$$S = \int \sqrt{-g}\left[-I^2(\phi, f(R))\left(\frac{1}{4}F^{\mu\nu}F_{\mu\nu} - \frac{\gamma_g}{8}\epsilon^{\mu\nu\alpha\beta}F_{\mu\nu}F_{\alpha\beta}\right) - RA^2 - \frac{\beta}{4m}\epsilon^{\mu\nu\alpha\beta}F_{\mu\nu}F_{\alpha\beta} - D_\mu\psi(D^\mu\psi)^*\right]d^4x - \int \sqrt{-g}\left[\frac{1}{2}g^{\mu\nu}\partial_\mu\phi\partial_\nu\phi + V(\phi)\right]d^4x. \quad (3)$$

This action usually contains the standard EM terms coupled to scalar fields (ϕ) like the inflaton or dilaton, being $V(\phi)$ its potential [11, 15]; coupling to curvature invariants (R) or a particular class of $f(R)$ theories [16]; coupling to a pseudo-scalar field like the axion (β) with a mass scale m [17], charged scalar fields (ψ) [18], and the presence of a constant γ_g that leads to a magnetic field with a net helicity [13]; here $\epsilon^{\mu\nu\alpha\beta}$ is totally antisymmetric tensor in four dimensions with $\epsilon^{0123} = (-g)^{-1/2}$. It is well known that, to create magnetic fields during inflation, the conformal invariance of the standard electrodynamics must be broken. One of the first models of inflamagnetogenesis was introduced by Ratra [19], where he proposed a conformal-breaking coupling between the scalar field (the inflaton) and the electromagnetic field. Many other mechanisms have been proposed following the same philosophy, and several conditions were obtained in order to explain the observed large-scale magnetic fields. However, serious obstacles arise in those mechanisms such as the strong-coupling problem where the theory becomes uncontrollable [20]; the backreaction problem in which an overproduction of the electric fields spoils inflation [20], and the curvature perturbation problem that enunciates the generation of both scalar and tensor curvature perturbations from PMFs would yield results in conflict with CMB observations [21]. More complete treatments of this subject can be found in Refs. [2, 9, 22–24].

2.2 Cosmological phase transitions

In the early Universe, there have been at least two phase transitions: the cosmological QCD phase transition (~250 MeV) and electroweak phase transition (~125 GeV) [25]. If these are first-order transitions, the Universe goes through an out of equilibrium process that generates bubble nucleation. As the Universe cools below the critical temperature, bubbles nucleate and grow, the walls of these bubbles collide with the others generating turbulence, and then dynamo mechanism creates and amplifies magnetic fields from this violent process that are concentrated later in the bubble walls [1]. Calculations of generation of magnetic fields during QCD phase transitions have been carried out by several authors [26–28]. Some cosmological phase transitions could generate uncorrelated magnetic fields given by [29]

$$B_l = B_L\left(\frac{L}{l}\right)^{3/2}, \quad (4)$$

and as we can see, PMFs generated by these mechanisms lead to a coherence length of the field smaller than the Hubble scale at that epoch, and weaker fields on galactic scales are obtained. However, the presence of helical fields can undergo processes of inverse cascade that transfers power from small to large scales, and thus, the result will be strong fields on very large scales [28].

2.3 Harrison's mechanism

Other alternative for the production of PMFs arises during the radiation era in regions that have nonvanishing vorticity. The first attempt at such a model was done by Harrison [30]; there, magnetic fields are created through vorticity generated by the velocity difference in the fluids present. For a formal derivation of the mechanism, see Refs. [31–33]. At temperatures larger than the electron mass, the interactions among protons, electrons, and photons are strong, and they are locked together. This means that all the system has the same angular velocity and seed fields cannot be generated. For temperatures below $T < 230$ eV, electrons and photons are tightly coupled through Thomson scattering, while the coupling between protons and photons is weak in this stage. Protons and electrons are still tightly coupled through Coulomb scattering, and so, the photon fluid drags the protons in its motion. Therefore, the difference of mass between electrons and protons will lead to non-zero electron and proton fluid angular velocities that give rise to currents and magnetic fields. Matarrese et al. [31] found that for comoving scales of ~ 1 Mpc, the amplitude of PMFs generated via this mechanism is around of $\sim 10^{-29}$ G today. This value of the magnetic field generated by the differential rotational velocity of charged particles is much smaller than those signals observed in clusters of galaxies. Now, if an initial vorticity is present during this epoch, magnetic fields may serve as seed for explaining the galactic fields; however, in the early Universe, the vorticity decays rapidly due to the expansion of the Universe, and therefore, this mechanism cannot work efficiently [32].

3. Magnetic spectra and correlation functions

Two models have been proposed to model PMFs. The first one consists in describing PMFs as an homogeneous field such that B^2 is the local density of the field and where we must require an anisotropic background (like Bianchi VII) to allow the presence of this field. Comparing those models with CMB quadrupole data, Barrow et al. [34] reported an amplitude of PMFs of $B < 6.8 \times 10^{-9} \left(\Omega_m h^2\right)^{1/2}$ G, and there they used the most general flat and open anisotropic cosmologies containing expansion rate and three-curvature anisotropies. However, they found that PMFs amplitude constraints are stronger than those imposed by nucleosynthesis, and therefore, this description hardly agrees with other cosmological probes. On the other hand, PMFs can also be described by a stochastic test field where B^2 would be related to the average density of the field instead. This description does not break neither isotropy nor homogeneity of the background Universe; hence, this scheme allows to have a PMF model concordant to the current constraints. In consequence, we will consider a stochastic primordial magnetic field (PMF) generated in the very early Universe which could have been produced during inflation (noncausal field) or after inflation (causal field) throughout the chapter. The PMF power spectrum which is defined as the Fourier transform of the two-point correlation can be written as

$$\langle B_l(\mathbf{k}) B_m^*(\mathbf{k}') \rangle = (2\pi)^3 \delta^3(\mathbf{k} - \mathbf{k}') \left(P_{lm}(k) P_B(k) + i \epsilon_{lmn} \hat{k}^n P_H(k) \right), \tag{5}$$

where $P_{lm}(k) = \delta_{lm} - \hat{k}_l \hat{k}_m$ is a projector onto the transverse plane[1], ϵ_{lmn} is the 3D Levi-Civita tensor, and $P_B(k)$ and $P_H(k)$ are the symmetric/antisymmetric parts of

[1] This projector has the property $P_{lm} \hat{k}^m = 0$ with $\hat{k} = \frac{\mathbf{k}}{k}$.

the power spectrum and represent the magnetic field energy density and absolute value of the kinetic helicity, respectively [35]:

$$\langle B_i(\mathbf{k})B_i^*(\mathbf{k}')\rangle = 2(2\pi)^3\delta^3(\mathbf{k}-\mathbf{k}')P_B(k), \tag{6}$$

$$-i\left\langle \epsilon_{ijl}\hat{k}^l B_i(\mathbf{k})B_j^*(\mathbf{k}')\right\rangle = 2(2\pi)^3\delta^3(\mathbf{k}-\mathbf{k}')P_H(k). \tag{7}$$

We assume that power spectrum scales as a simple power law

$$P_B(k) = A_B k^{n_B}, \quad P_H(k) = A_H k^{n_H}. \tag{8}$$

We usually parametrize the fields through a convolution with a 3D-Gaussian window function smoothed over a sphere of comoving radius λ, $B_i(k) \to B_i(k)\times f(k)$, with $f(k) = e^{(-\lambda^2k^2/2)}$ [4]. We also define B_λ as the comoving PMF strength scaled to the present day on λ:

$$\begin{aligned}\langle B^i(\mathbf{x})B_i(\mathbf{x})\rangle|_\lambda \equiv B_\lambda^2 &= \frac{1}{(2\pi)^6}\int\int d^3k d^3k' e^{(-i\mathbf{x}\cdot\mathbf{k}+i\mathbf{x}\cdot\mathbf{k}')}\langle B^i(\mathbf{k})B_i^*(\mathbf{k}')\rangle|f(k)|^2,\\ &= \frac{A_B}{(2\pi)^2}\frac{2}{\lambda^{n_B+3}}\Gamma\left(\frac{n_B+3}{2}\right),\end{aligned} \tag{9}$$

and we define $\mathcal{B}_\lambda$ as the comoving kinetic helical PMF strength scaled to the present day on λ:

$$\begin{aligned}\langle(\nabla\times\mathbf{B}(\mathbf{x}))_i B^i(\mathbf{x})\rangle|_\lambda &\equiv \mathcal{B}_\lambda^2\\ &= \frac{i\epsilon_{ilj}}{(2\pi)^6}\int\int d^3k d^3k' e^{(-i\mathbf{x}\cdot\mathbf{k}+i\mathbf{x}\cdot\mathbf{k}')}\langle k^j B^l(\mathbf{k})B_i^*(\mathbf{k}')\rangle|f(k)|^2,\\ &= \frac{|A_H|}{(2\pi)^2}\frac{2}{\lambda^{n_H+4}}\Gamma\left(\frac{n_H+4}{2}\right),\end{aligned} \tag{10}$$

with Γ being the gamma function. Then, we obtain the amplitudes as follows:

$$A_B = \frac{B_\lambda^2 2\pi^2\lambda^{n_B+3}}{\Gamma\left(\frac{n_B+3}{2}\right)}, \quad A_H = \frac{H_\lambda^2 2\pi^2\lambda^{n_H+3}}{\Gamma\left(\frac{n_H+4}{2}\right)}, \quad \text{with} \quad n_B > -3, n_H > -4. \tag{11}$$

The most general case of the power spectrum for magnetic fields can be studied, if we assume that it is non-zero for $k_m \le k \le k_D$, being k_m an infrared cutoff and k_D an ultraviolet cutoff corresponding to damping scale of the field written as [4]

$$k_D \approx (1.7\times10^2)^{\frac{2}{n+5}}\left(\frac{B_\lambda}{10^{-9}nG}\right)^{\frac{-2}{n+5}}\left(\frac{k_\lambda}{1Mpc^{-1}}\right)^{\frac{n+3}{n+5}} h^{\frac{1}{n+5}}\frac{1}{Mpc}. \tag{12}$$

Hereafter we simply set this scale at $k_D \sim \mathcal{O}(10)\text{Mpc}^{-1}$ [4]. Given the Schwarz inequality [36],

$$\lim_{k'\to k}\langle\mathbf{B}(k)\cdot\mathbf{B}^*(k')\rangle \ge |\lim_{k'\to k}\left\langle\left(\hat{\mathbf{k}}\times\mathbf{B}(k)\right)\cdot\mathbf{B}^*(k')\right\rangle|, \tag{13}$$

an additional constraint is found for these fields

$$|A_H| \le A_B k^{n_B-n_H}. \tag{14}$$

In the case where $A_H = A_B$ and $n_B = n_H$, we define the maximal helicity condition. We will also parametrize the infrared cutoff by a single constant parameter α:

$$k_m = \alpha k_D, \quad 0 \leq \alpha < 1 \tag{15}$$

which in the case of inflationary scenarios would correspond to the wave mode that exits the horizon at inflation epoch and, for causal modes, would be important when this scale is larger than the wave number of interest (as claimed by Kim et al. [37]). Thus, this infrared cutoff would be important in order to constrain PMF parameters and magnetogenesis models [37–40]. Eq. (15) gives only a useful mathematical representation to constrain these cutoff values via cosmological datasets (for this case, the parameter space would be given by $(\alpha, k_D, B_\lambda, H_\lambda, n_H, n_B)$), and therefore we want to point out that the latter expression does not state any physical relation between both wave numbers. In [38, 39], they showed constraints on the maximum wave number k_D as a function of n_B via big bang nucleosynthesis (BBN), and they considered the maximum and minimum wave numbers as independent parameters. In fact, in [3] they found out that the integration scheme used for calculating the spectrum and bispectrum of PMFs is exactly the same if we parametrize k_m as seen in (15) or if we consider $(k_m, k_D, B_\lambda, H_\lambda, n_H, n_B)$ as independent parameters.

Thus the inclusion of k_m is done only for studying at a phenomenological level, and its effects on the CMB are shown in more detail in [3, 41]. At background level, we need only the energy density of the PMF which is given by $\rho_B = \langle B^2(\mathbf{x})\rangle/(8\pi)$; therefore, by using Eqs. (8) and (9), we get (for the spatial dependence)

$$\rho_B = \frac{\langle B^2(\mathbf{x})\rangle}{8\pi} = \frac{2}{8\pi}\int_{k_m}^{k_D} d^3k P_B(k) = \frac{\lambda^{n_B+3}}{8\pi}\frac{B_\lambda^2}{\Gamma\left(\frac{n_B+5}{2}\right)}\left[k_D^{n_B+3} - k_m^{n_B+3}\right]. \tag{16}$$

Here, only the non-helical term contributes to the energy density of the PMF in the Universe. In Ref. [38], this equation is also reported, and we will study in more detail their effects on the CMB later. In order to study the impact of PMFs on cosmological perturbations, we start writing the magnetic energy momentum tensor (EMT)

$$T_0^0 = \frac{-1}{8\pi a^4}|B(\mathbf{x})|^2, \quad T_j^i = \frac{1}{4\pi a^4}\left(\delta_j^i \frac{|B(\mathbf{x})|^2}{2} - B_j(\mathbf{x})B^i(\mathbf{x})\right), \quad T_i^0 = 0, \tag{17}$$

where we can see that EMT of PMFs is quadratic in the fields [42]. Due to the high conductivity in the primordial Universe, the electric field is suppressed, and the magnetic one is frozen into the plasma, and consequently we have that $B_i(\mathbf{x}, \tau) = B_i(\mathbf{x})a^{-2}(\tau)$. Then, the spatial part of magnetic field EMT in Fourier space is given by

$$T_j^i(\mathbf{k}, \tau) = \frac{-1}{32\pi^4 a^4}\int d^3k' \left[B^i(\mathbf{k}')B_j(\mathbf{k}-\mathbf{k}') - \frac{1}{2}\delta_j^i B^l(\mathbf{k}')B_l(\mathbf{k}-\mathbf{k}')\right], \tag{18}$$

and the two-point correlation tensor related to the spatial dependence (18) gives

$$\begin{aligned}\langle T_{ij}(\mathbf{k})T_{lm}^*(\mathbf{p})\rangle = \frac{1}{1024\pi^8}\int\int d^3p' d^3k' \langle B_i(\mathbf{k}')B_j(\mathbf{k}-\mathbf{k}')B_l^*(\mathbf{p}')B_m^*(\mathbf{p}-\mathbf{p}')\rangle \\ + \ldots\langle\ldots\rangle_{lm}\delta_{ij} + \ldots\langle\ldots\rangle_{ij}\delta_{lm} + \ldots\delta_{ij}\delta_{lm},\end{aligned} \tag{19}$$

where we can apply the Wick theorem because the stochastic fields are Gaussianly distributed

$$\begin{aligned}\langle B_i(\mathbf{k}')B_j(\mathbf{k}-\mathbf{k}')B_l^*(\mathbf{p}')B_m^*(\mathbf{p}-\mathbf{p}')\rangle &= \langle B_i(\mathbf{k}')B_j(\mathbf{k}-\mathbf{k}')\rangle\langle B_l^*(\mathbf{p}')B_m^*(\mathbf{p}-\mathbf{p}')\rangle \\ &+ \langle B_i(\mathbf{k}')B_l^*(\mathbf{p}')\rangle\langle B_j(\mathbf{k}-\mathbf{k}')B_m^*(\mathbf{p}-\mathbf{p}')\rangle \\ &+ \langle B_i(\mathbf{k}')B_m^*(\mathbf{p}-\mathbf{p}')\rangle\langle B_l^*(\mathbf{p}')B_j(\mathbf{k}-\mathbf{k}')\rangle.\end{aligned} \tag{20}$$

On the other hand, the equations for the adimensional energy density of magnetic field and spatial part of the electromagnetic energy momentum tensor respectively written in Fourier space are given as

$$\begin{aligned}\rho_B(\mathbf{k}) &\equiv \frac{1}{8\pi\rho_{\gamma,0}}\int\frac{d^3p}{(2\pi)^3}B_l(\mathbf{p})B^l(\mathbf{k}-\mathbf{p}), \\ \Pi_{ij}(\mathbf{k}) &\equiv \frac{1}{4\pi\rho_{\gamma,0}}\int\frac{d^3p}{(2\pi)^3}\left[\frac{\delta_{ij}}{2}B_l(\mathbf{p})B^l(\mathbf{k}-\mathbf{p}) - B_i(\mathbf{p})B_j(\mathbf{k}-\mathbf{p})\right],\end{aligned} \tag{21}$$

where we express each component of the energy momentum tensor in terms of photon energy density $\rho_\gamma = \rho_{\gamma,0}a^{-4}$, with $\rho_{\gamma,0}$ being its present value[2]. We can also see that using the previous definition, the EMT can be written as $T^i_j(\mathbf{k},\tau) \equiv \rho_\gamma(\tau)\Pi^i_j(\mathbf{k})$. Since the spatial EMT is symmetric, we can decompose this tensor into two scalars (ρ_B, $\Pi^{(S)}$), one vector ($\Pi_i^{(V)}$), and one tensor ($\Pi_{ij}^{(T)}$) components:

$$\Pi_{ij} = \frac{1}{3}\delta_{ij}\rho_B + \left(\hat{k}_i\hat{k}_j - \frac{1}{3}\delta_{ij}\right)\Pi^{(S)} + \left(\hat{k}_i\Pi_j^{(V)} + \hat{k}_j\Pi_i^{(V)}\right) + \Pi_{ij}^{(T)} \tag{22}$$

which obey to $\hat{k}^i\Pi_i^{(V)} = \hat{k}^i\Pi_{ij}^{(T)} = \Pi_{ii}^{(T)} = 0$ [45, 46]. The components of this tensor are recovered by applying projector operators defined as

$$\begin{aligned}\rho_B &= \delta^{ij}\Pi_{ij} \\ \Pi^{(S)} &= \left(\delta^{ij} - \frac{3}{2}P^{ij}\right)\Pi_{ij} = \mathcal{P}^{ij}\Pi_{ij} \\ \Pi_i^{(V)} &= \hat{k}^{(j}P_i^{l)}\Pi_{lj} = \mathcal{Q}_i^{jl}\Pi_{lj} \\ \Pi_{ij}^{(T)} &= \left(P_i^{(a}P_j^{b)} - \frac{1}{2}P^{ab}P_{ij}\right)\Pi_{ab} = \mathcal{P}_{ij}^{ab}\Pi_{ab},\end{aligned} \tag{23}$$

where (..) in the indices denotes symmetrization [47]. The two-point correlation tensor related to Eq. (21) is

---

[2] The dimensional energy density of magnetic field showed here is written with different notation in [43] $\Omega_B \equiv \frac{B^2}{8\pi a^4\rho_\gamma}$ and in [7, 44] $\Delta_B \equiv \frac{B^2}{8\pi a^4\rho_\gamma}$.

$$
\begin{aligned}
\langle \Pi_{ij}(\mathbf{k})\Pi^{*}_{lm}(\mathbf{p})\rangle = \frac{1}{(4\pi\rho_{\gamma,0})^{2}}\delta^{(3)}(\mathbf{k}-\mathbf{p})\int d^{3}k'\Big[\Big(&P_B(k')P_B(|\mathbf{k}-\mathbf{k}'|)P_{il}(k')P_{jm}(|\mathbf{k}-\mathbf{k}'|) \\
&- P_H(k')P_H(|\mathbf{k}-\mathbf{k}'|)\epsilon_{ilt}\epsilon_{jms}\hat{k}'_t\left(\widehat{\mathbf{k}-\mathbf{k}'}\right)_s \\
&+ iP_B(k')P_H(|\mathbf{k}-\mathbf{k}'|)P_{il}(k')\epsilon_{jmt}\left(\widehat{\mathbf{k}-\mathbf{k}'}\right)_t \\
&+ iP_B(k')P_H(|\mathbf{k}-\mathbf{k}'|)\epsilon_{ilt}P_{jm}(|\mathbf{k}-\mathbf{k}'|)\hat{k}'_t + (l\leftrightarrow m)\Big) \\
&+ ...\delta_{ij} + ...\delta_{lm} + ...\delta_{ij}\delta_{lm}\Big],
\end{aligned}
\tag{24}
$$

where we use Eqs. (5) and (20). In this work, we are only focused on the scalar mode of the PMFs. To determine the effect on cosmic perturbations, it is necessary to compute the scalar correlation functions of PMFs using the projector operators:

$$
\begin{aligned}
\langle \rho_B(\mathbf{k})\rho_B^{*}(\mathbf{k}')\rangle &= \delta^{ij}\delta^{lm}\langle \Pi_{ij}(\mathbf{k})\Pi^{*}_{lm}(\mathbf{k}')\rangle, \\
\langle \Pi^{(S)}(\mathbf{k})\Pi^{(S)*}(\mathbf{k}')\rangle &= \mathcal{P}^{ij}\mathcal{P}^{lm}\langle \Pi_{ij}(\mathbf{k})\Pi^{*}_{lm}(\mathbf{k}')\rangle.
\end{aligned}
\tag{25}
$$

These convolutions can be written in terms of spectra as follows [35, 48]:

$$
\langle \rho_B(\mathbf{k})\rho_B^{*}(\mathbf{k}')\rangle = (2\pi)^3|\rho(k)|^2\delta^{(3)}(\mathbf{k}-\mathbf{k}'), \tag{26}
$$

$$
\left\langle \Pi^{(S)}(\mathbf{k})\Pi^{(S)*}(\mathbf{k}')\right\rangle = (2\pi)^3\left|\Pi^{(S)}(k)\right|^2\delta^{(3)}(\mathbf{k}-\mathbf{k}'). \tag{27}
$$

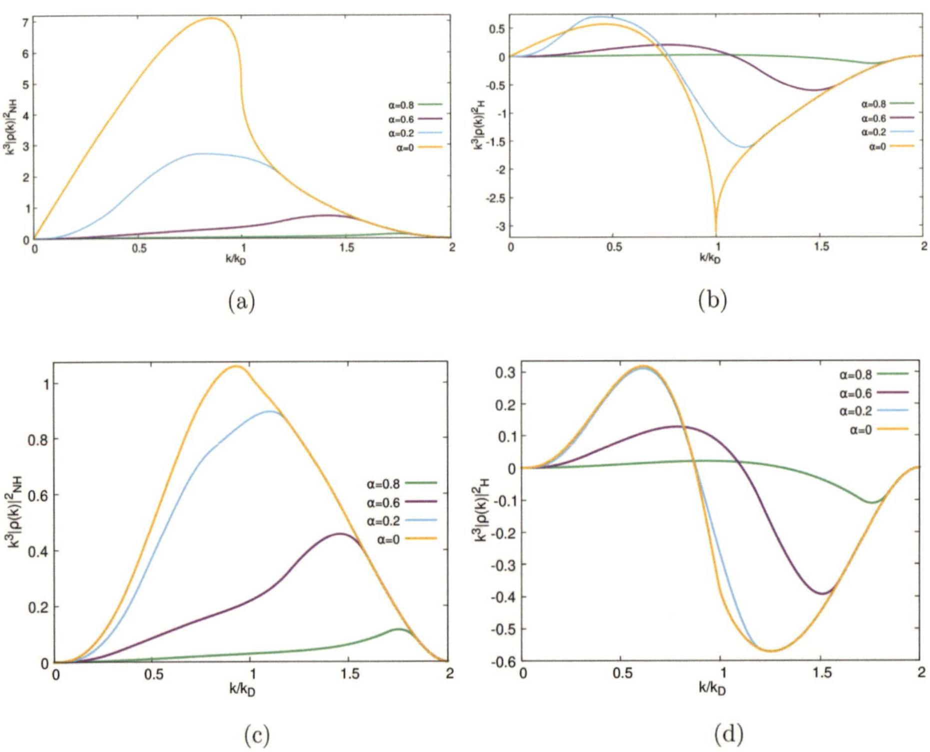

Figure 1.
Non-helical contribution to $k^3|\rho(k)|^2$ for different spectral indices in units of $A^2_{(B),H}/\left(8(2\pi)^5\rho^2_{\gamma,0}\right)$ versus k/k_D. Here we show the effect of an IR cutoff parametrized with α on the magnetic power spectrum. (a) Non-helical contribution to $k^3|\rho(k)|^2$ for $n_B = n_H = -5/2$. (b) Helical contribution to $k^3|\rho(k)|^2$ for $n_B = n_H = -5/2$. (c) Non-helical contribution to $k^3|\rho(k)|^2$ for $n_B = n_H = -3/2$. (d) Helical contribution to $k^3|\rho(k)|^2$ for $n_B = n_H = -3/2$.

Thus, using Eqs. (24)–(27), along with the Wick's theorem (20), the spectra take the form

$$|\rho(k)|^2 = \frac{1}{8(2\pi)^5\rho_{\gamma,0}^2}\int d^3p'\left((1+\mu^2)P_B(p')P_B(|\mathbf{k}-\mathbf{p}'|) - 2\mu P_H(p')P_H(|\mathbf{k}-\mathbf{p}'|)\right), \tag{28}$$

$$\left|\Pi^{(S)}(k)\right|^2 = \frac{1}{8(2\pi)^5\rho_{\gamma,0}^2}\int d^3p'\left(\left[4-3\gamma^2+\beta^2(-3+9\gamma^2)-6\beta\gamma\mu+\mu^2\right]P_B(p')\right.$$
$$\left. P_B(|\mathbf{k}-\mathbf{p}'|) - (6\beta\gamma-4\mu)P_H(p')P_H(|\mathbf{k}-\mathbf{p}'|)\right), \tag{29}$$

where the angular functions are defined as

$$\beta = \frac{\mathbf{k}\cdot(\mathbf{k}-\mathbf{k}')}{k|\mathbf{k}-\mathbf{k}'|}, \quad \mu = \frac{\mathbf{k}'\cdot(\mathbf{k}-\mathbf{k}')}{k'|\mathbf{k}-\mathbf{k}'|}, \quad \gamma = \frac{\mathbf{k}\cdot\mathbf{k}'}{kk'}. \tag{30}$$

The above relations and properties were obtained using the xAct software [49], and they agree with those reported in [35, 47]. Given these results, we are able to analyze the effects of PMFs on CMB by adding the previous contributions to the

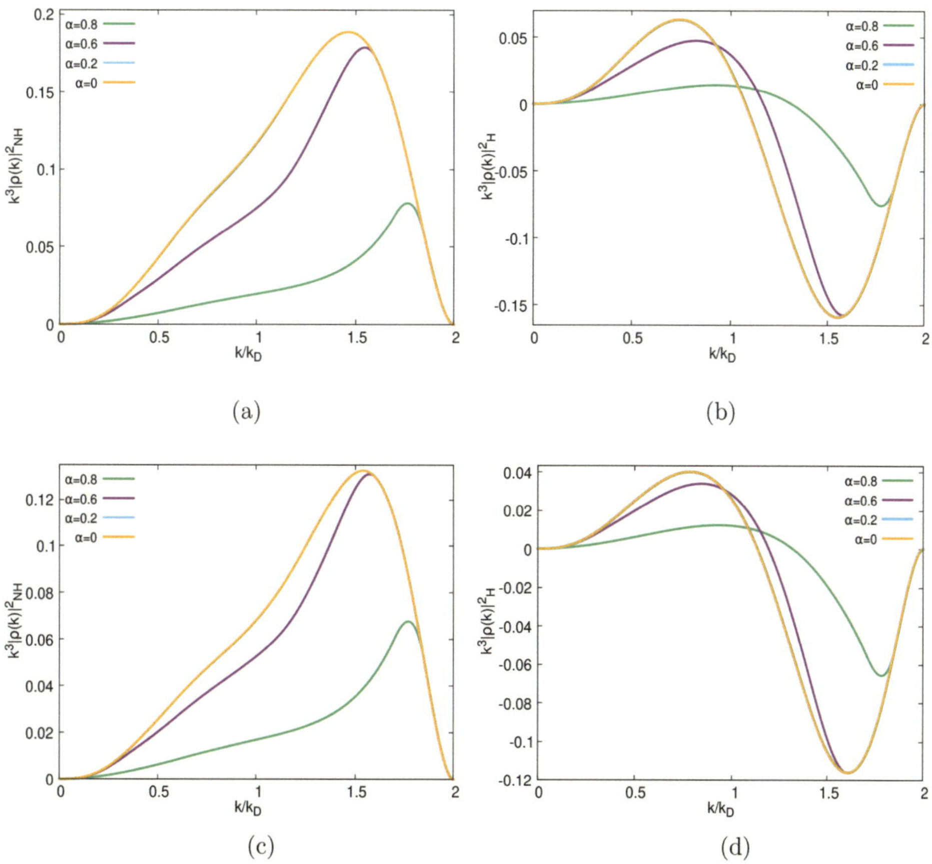

Figure 2.
Non-helical contribution to $k^3|\rho(k)|^2$ for different spectral indices in units of $A^2_{(B),H}/\left(8(2\pi)^5\rho^2_{\gamma,0}\right)$ versus k/k_D. Here we show the effect of an IR cutoff parametrized with α on the magnetic power spectrum. (a) Non-helical contribution to $k^3|\rho(k)|^2$ for $n_B = n_H = 1$. (b) Helical contribution to $k^3|\rho(k)|^2$ for $n_B = n_H = 1$. (c) Non-helical contribution to $k^3|\rho(k)|^2$ for $n_B = n_H = 2$. (d) Helical contribution to $k^3|\rho(k)|^2$ for $n_B = n_H = 2$.

CMB angular power spectrum. Indeed, some authors [42, 50–53] have added the above spectrum relations in Boltzmann codes like CAMB [54] or CMBeasy [55], while other authors [4, 5, 56, 57] have analyzed the effects of these fields through approximate solutions.

Using the integration scheme for the Fourier spectra reported in [41], we obtain the solution for the magnetic spectra for different contributions. In **Figures 1** and 2, we show the total contribution for $k^3|\rho(k)|^2$ in the maximal helical case for several spectral indices and values of α. Here we can see that for $n_B < 0$, the spectrum is red while for $n > 0$ the biggest contribution comes from large wave numbers. In **Figure 3**, the scalar part of the anisotropic stress and the effect of an IR cutoff on its spectrum are displayed.

4. Effects of the background PMFs on the CMB

The presence of energy density of the background PMF increases total radiation-like energy density ρ_r and modifies the standard dynamics of the background

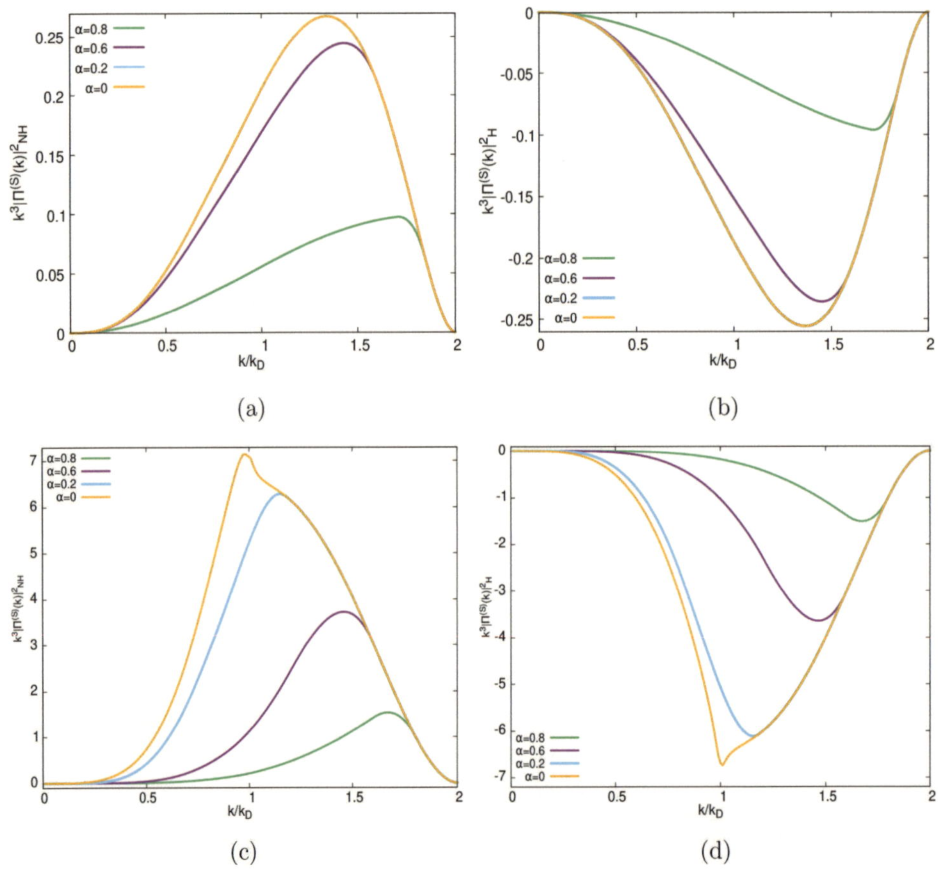

Figure 3.

Non-helical contribution to $k^3\left|\Pi(k)^{(S)}\right|^2$ *for different spectral indices in units of* $A^2_{(B),H}/\left(8(2\pi)^5\rho^2_{\gamma,0}\right)$ *versus* k/k_D. *Here we show the effect of an IR cutoff parametrized with* α *on the magnetic power spectrum. (a) Non-helical contribution to* $k^3\left|\Pi(k)^{(S)}\right|^2$ *for* $n_B = n_H = 2$. *(b) Helical contribution to* $k^3\left|\Pi(k)^{(S)}\right|^2$ *for* $n_B = n_H = 2$. *(c) Non-helical contribution to* $k^3\left|\Pi(k)^{(S)}\right|^2$ *for* $n_B = n_H = -5/2$. *(d) Helical contribution to* $k^3\left|\Pi(k)^{(S)}\right|^2$ *for* $n_B = n_H = -5/2$.

Universe producing considerable effects on the primary temperature fluctuations of the CMB. In this section, we will study the effects of background PMF on the CMB following the early work discussed in Ref. [38]. First, the total energy density and pressure are now written as $\rho = \rho^{(\Lambda CDM)} + \rho_B$ and $P = \sum_i w_i \rho_i^{(\Lambda CDM)} + P_B$, respectively (being (ΛCDM) the components of matter in the standard model of cosmology), modifying the solution of the Friedmann's equation:

$$\left(\frac{a'}{a}\right)^2 = \frac{8\pi G}{3} a^2 \rho, \quad \left(\frac{a'}{a}\right)' = -\frac{4\pi G}{3} a^2 (\rho + 3P), \tag{31}$$

where G is the gravitational constant and $a' \equiv da/d\tau$ with τ the conformal time. In order to study the effects of PMFs on the CMB, we include a background magnetic density given by Eq. (16) into the Boltzmann code CLASS [58]. As first shown by Ref. [38], the speed of sound in baryon fluid is described by $c_{s,eff}^2 = c_s^2 + \rho_B/\rho_t$, where c_s^2 is the speed of sound without magnetic field [59]

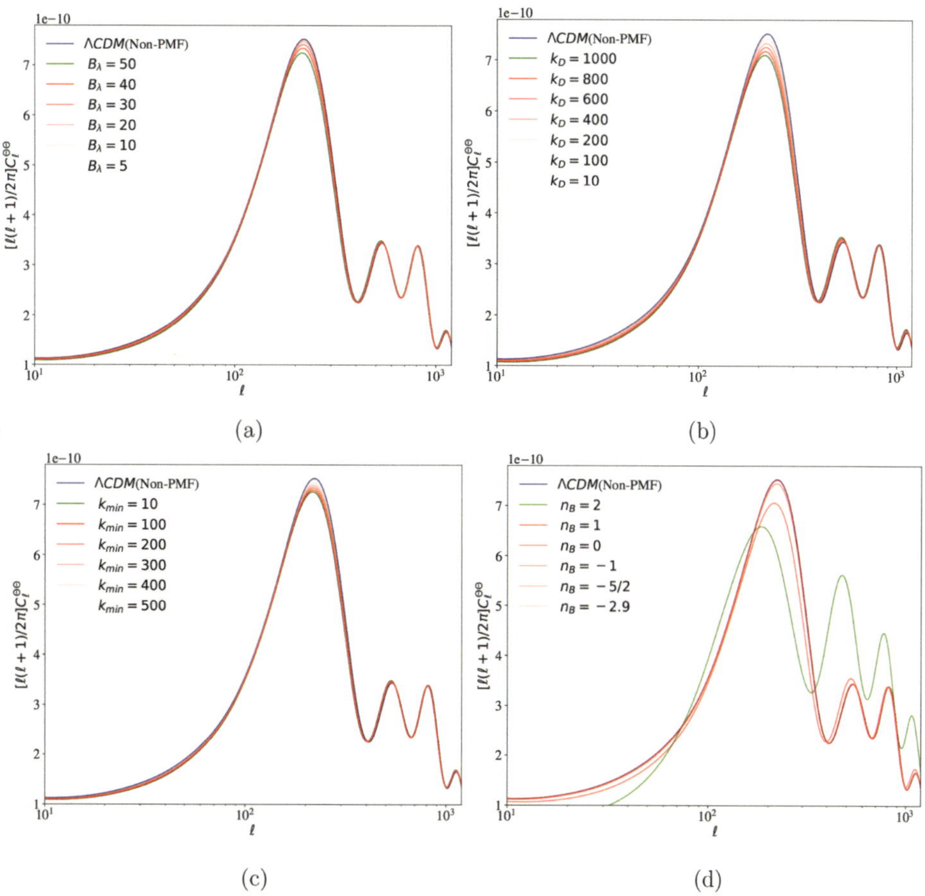

Figure 4.
Spectrum of CMB temperature anisotropies with PMFs obtained numerically from CLASS code. Each plot displays the effect of B_λ (a), k_D (b), k_{min} (c), and n_B (d) on the CMB spectrum. Here the blue line stands for the model without PMF, and B_λ is in units of nG, and k in units of Mpc^{-1}. (a) $l(l+1)C_l$ with $k_D = 100$, $k_{min} = 0$, $n_B = -2$. The green line assumes $\rho_B/\rho_\gamma = 0.0041$. (b) $l(l+1)C_l$ with $B_\lambda = 20$ nG, $k_{min} = 0$, $n_B = -2$. The green line assumes $\rho_B/\rho_\gamma = 0.0065$. (c) $l(l+1)C_l$ with $B_\lambda = 20$ nG, $k_D = 200$, $n_B = -2$. The green line assumes $\rho_B/\rho_\gamma = 0.0038$. (d) $l(l+1)C_l$ with $B_\lambda = 20$ nG, $k_{min} = 0$, $k_D = 400$. The green line assumes $\rho_B/\rho_\gamma = 0.0044$

$$c_s^2 = \frac{1}{3(1+R)}, \quad \text{with}: \ R \equiv 3\rho_b/4\rho_r, \tag{32}$$

and ρ_t is the total energy density. Also, including this modification in the thermodynamic structure in the CLASS code, we obtain the spectrum of CMB temperature anisotropies shown in **Figure 4**. Here we can observe the effects of different parameters enclosed in Eq. (16) on the CMB spectrum. With a PMF in the primordial plasma, the time of matter-radiation density equality ($\rho_m = \rho_r$) increases enhancing the amplitude of all peaks because there is not enough time to be suppressed for the cosmic expansion. However, the contrast between odd and even peaks is reduced because it further depends on (ρ_m/ρ_r) corresponding to the balance between gravity and the total radiation pressure [60]. Secondly, an important effect of PMFs comes from increasing in sound speed c_s. In fact, the peak location depends on the angle $\theta = d_s(\tau_{dec})/d_A(\tau_{dec})$, where $d_s(\tau_{dec})$ is the physical sound horizon at decoupling and $d_A(\tau_{dec})$ is the angular diameter distance at decoupling [60]

$$d_s = a \int_{\tau_{ini}}^{\tau} c_s d\tau, \quad d_A = \frac{1}{1+z}\int_0^z \frac{dz'}{H(z')}. \tag{33}$$

The angular diameter distance depends on the late history after decoupling (Ω_Λ, h), whereas physical sound horizon is further affected by the value of c_s. By adding PMFs to the primordial plasma, we increase the effective speed of sound which in turn increases the angle of the location of peaks, boosting the peaks to small $l's$ (this can be understood geometrically using $\theta \sim \pi/l$), i.e., shifting the acoustic peaks to the left as we see in **Figure 5a**. Finally, since the value of the total radiation energy density is larger with PMFs, the gravitational potentials ϕ, ψ decay more quickly after their wavelengths become smaller than the sound horizon (see **Figure 5a**).

In summary, accounting for a background PMF in our model modifies the shape of the temperature power spectrum significantly for large multipolar numbers, that

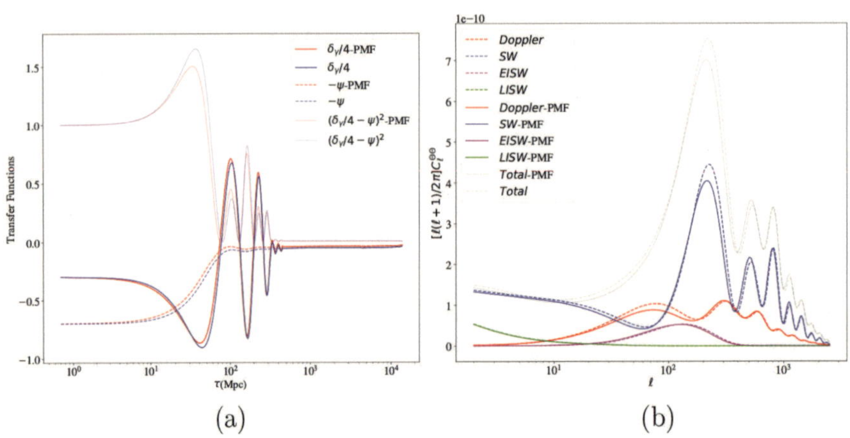

Figure 5.
A snapshot of the transfer functions and $l(l+1)C_l$ temperature accounting for a PMF with $\rho_B/\rho_\gamma = 0.0041$. Left panel we plot the numerical solution obtained from CLASS code for δ_γ and ψ with and without PMFs. Right panel we plot the CMB spectrum with and without PMFs showing the individual contributions explained in the text: Sachs-Wolfe (SW), Doppler, early integrated Sachs-Wolfe (EISW), and late integrated Sachs-Wolfe (LISW). Note that the total spectrum labeled by the black line corresponds to all correlations. (a) Effect of PMF on the transfer functions for k = 0.1 *Mpc^{-1}. (b) Individual contribution of $l(l+1)C_l$ with PMFs.*

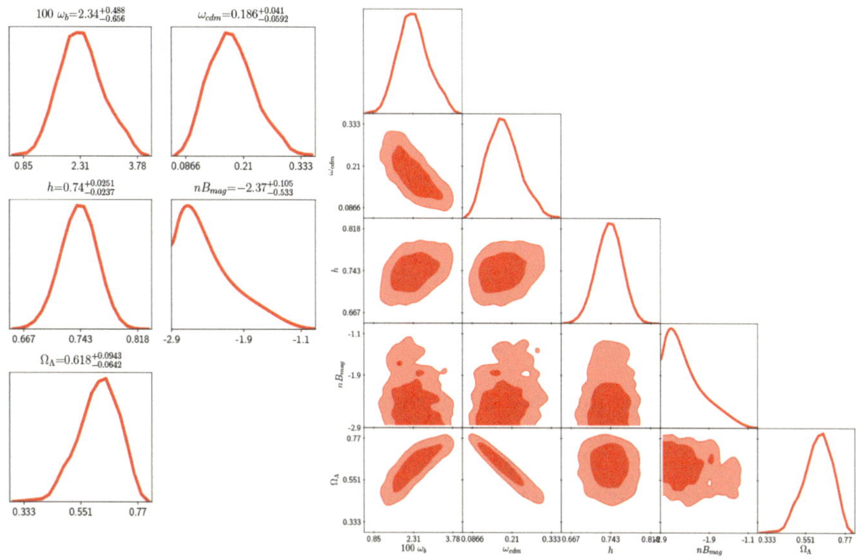

(a) (b)

Figure 6.
Left: Results of MCMC constrained with BAO data 2015 and with Planck simulated data 2013. Right: Triangle plot of the results of MCMC. Regions are the 68% and 95% confidence level. Here we use B = 1nG, k_D = 100 Mpc^{-1}, and $k_{min} = 0$. (a) Bidimensional plot of the results of the MCMC analysis with PMFs. (b) Triangle plot of the results of the MCMC analysis with PMFs.

is, the Sachs-Wolfe (SW), Doppler, and early integrated Sachs-Wolfe (EISW) contributions are quite affected by the magnetic field. This fact can be noticed in **Figure 5b**, where we plot the features of PMFs ($\rho_B/\rho_\gamma = 0.0041$) for several contributions of the CMB spectrum. Since the late integrated Sachs-Wolfe (LISW) comes from interactions of the photons after last scattering, PMFs do not play a sizable role in this contribution. On the other hand, the EISW signal is shifted to small $l's$ because modes related to ψ, ϕ entered to sub-horizon scales earlier than if they had done without PMFs. This boost is also seen in the SW where the acoustic peak positions are shifted to larger scales. For $l > 100$, odd Doppler peaks are enhanced with respect to the ratio of baryon and radiation content [60]; hence, PMFs produce suppression in the amplitude for odd peaks, while the even ones remain unaltered. These features are illustrated in **Figure 5b**.

In **Figure 6** we show the bidimensional and triangle plots of the MCMC with one magnetic parameter and some of the ΛCDM model. We derived the constraints on the spectral index of PMF: B = 1nG, k_D = 100 Mpc^{-1}, and $-3 < n_B < 0$ at 95% confidence level. Therefore, cosmological datasets (BAO data 2015 and with Planck simulated data 2013) strongly favor invariant scale fields driven by inflationary scenarios. In order to derive the constraints with current data on this kind of PMFs, we performed a MCMC analysis using the Monte Python code [61].

5. Magnetic contribution to CMB anisotropies

Since PMFs affect the evolution of cosmological perturbations, these fields might leave significant signals on the CMB. Basically, PMFs add three contributions to the temperature and polarization of the CMB spectra, such as the scalar, vector, and tensor, which have been deeply studied [4, 56, 62, 63]. For the scalar contribution, the shape of the temperature anisotropy (TT mode) presents an increase on large scales, and it also shifts the acoustic peaks via fast magnetosonic waves; nevertheless the main effect of the scalar mode lies on large multipolar numbers, since the primary CMB is significantly suppressed by the Silk damping in these scales [4, 5]. Next, the vector contribution leaves an indistinguishable signal, because in standard cosmology, vector contributions decay with time and do not affect the CMB anisotropies considerably [4]. Further, vector mode peaks where primary CMB is suppressed by Silk damping and so dominates over the scalar ones in small scales [64]. Vector modes are also very interesting in the polarization spectra; in particular, they induce B modes with amplitudes slightly larger than any other contribution, allowing us to constrain better PMFs in the next CMB polarization experiments [52].

Finally, tensor modes induce gravitational wave perturbations that lead to CMB temperature and polarization anisotropies on large angular scales, and the passive tensor modes (produced by the presence of PMFs before neutrino decoupling) generate the most significant magnetic contributions, so those modes become relevant to study the nature of PMFs [48, 65, 66]. Moreover, if helical PMFs are presented before recombination, they affect drastically the parity-odd CMB cross correlations implying a strong feature of parity violation in the early Universe [50]. Using the total angular momentum formalism introduced by [67], the angular power spectrum of the CMB temperature anisotropy is given as

$$(2l+1)^2 C_l^{\mathcal{X}\mathcal{X}} = \frac{2}{\pi}\int \frac{dk}{k} \sum_{m=-2}^{2} k^3 \mathcal{X}_l^{(m)*}(\tau_0,k)\mathcal{X}_l^{(m)}(\tau_0,k), \tag{34}$$

where $m = 0, \pm 1, \pm 2$ are the scalar, vector, and tensor perturbation modes and $\mathcal{X} = \{\Theta, E, B\}$. Here $\Theta_l^{(m)}(\tau_0,k)$ are the temperature fluctuation $\frac{\delta T}{T}$ multipolar moments, and B, E represent the polarization of electric and magnetic type, respectively. In large scales, one can neglect the contribution on CMB temperature anisotropies by ISW effect in the presence of a PMF [4]. Therefore, considering just the fluctuation via PMF perturbation, the temperature anisotropy multipole moment for $m = 0$ becomes [4]

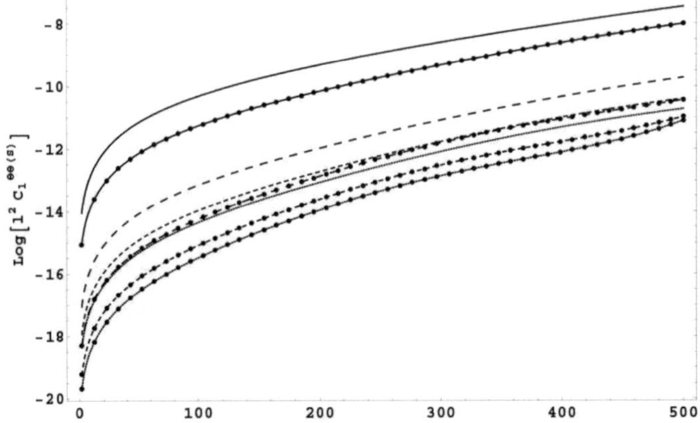

Figure 7.
Plot of the CMB temperature power spectrum induced by scalar magnetic perturbations, where the lines with filled circles are for $n = 2$ and the other ones for $n = 5/2$. Here, the solid lines refer to $B_\lambda = 10$ nG, large dashed lines for $B_\lambda = 8$ nG, small dashed lines refer to $B_\lambda = 5$ nG, and dotted lines for $B_\lambda = 1$ nG. Figure taken from [41].

$$\frac{\Theta_l^{(S)}(\tau_0,k)}{2l+1} \approx \frac{-8\pi G}{3k^2 a_{dec}^2}\rho_B(\tau_0,k) j_l(k\tau_0), \tag{35}$$

where a_{dec} is the value of scalar factor at decoupling, G is the gravitational constant, and j_l is the spherical Bessel function. Substituting the last expression in (34) with $\mathcal{X} = \Theta$, the CMB temperature anisotropy angular power spectrum is given by

$$l^2 C_l^{\Theta\Theta(S)} = \frac{2}{\pi}\left(\frac{8\pi G}{3a_{dec}^2}\right)^2 \int_0^\infty \frac{|\rho_B(\tau_0,k)|^2}{k^2} j_l^2(k\tau_0) l^2 dk. \tag{36}$$

Here, for our case, we should integrate only up to $2k_D$ since it is the range where energy density power spectrum is not zero. The result of the angular power spectrum induced by scalar magnetic perturbations given by (36) is shown in **Figure** 7. There, we plot the $\log l^2 C_l^{\Theta\Theta}$ in order to compare our results with those found by [4]. We calculate the angular power spectrum of CMB in units of $\frac{2}{\pi}\left(\frac{8\pi G}{3a_{dec}^2}\right)^2$. One of the important features of the CMB power spectrum (scalar mode) with PMFs is that the distortion is proportional to strength of PMF and decreases with the spectral index and we must expect its greatest contribution at low multipoles.

5.1 Infrared cutoff in the CMB spectra

Studying the effect of this lower cutoff of CMB spectra, we can constrain PMF generation models. **Figure 8** shows the effects of PMFs on the scalar mode of CMB spectra. Here we did a comparison between the Cls with a null cutoff with respect to Cls generated by values of cutoff different from zero. The horizontal solid line shows the comparison with $k_m = 0$, $k_m = 0.001k_D$, and $k_m = 0.1k_D$; no difference in effectiveness was found between these values. The dashed lines report a significant difference of the Cls for values of $k_m = 0.3k_D$, $k_m = 0.7k_D$, and $k_m = 0.9k_D$. It is appropriate to remark that power spectrum of causal fields is a smooth function in the k-space without any sharp cutoff coming from the original mechanism; now, given the parametrization

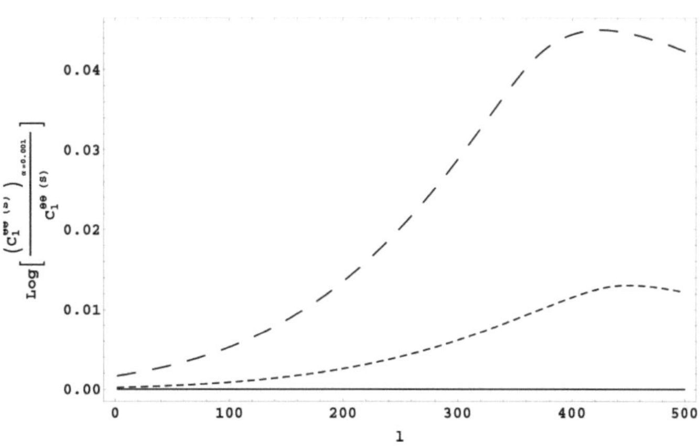

Figure 8.
Comparison between the CMB temperature power spectrum induced by scalar PMF at $k_m = 0.001k_D$ lower cutoff, with respect to the other ones with different values of infrared cutoff. Here, the solid horizontal line is for $k_m = 0.1k_D$; small and large dashed lines refer to $k_m = 0.3k_D$ and $k_m = 0.4k_D$, respectively. Figure taken from [41].

introduced here, we notice from **Figure** 7 in [41] that for α very small, the calculations agree with previous work. It can be thought as contribution of the super horizon modes is negligible, and one would expect that scales as $\sim k^4$, for instance. Also, one of the characteristics of this dependence is the existence of a peak; indeed, for large values of α, the peak moves to left as we see, for instance, with $\alpha = 0.4$ where the peak is in $l \sim 380$ while for $\alpha = 0.9$ the peak is shifted to $l \sim 200$.

6. Conclusions

In this chapter, we worked on the assumption that in the early Universe, a weak magnetic field was created. This PMF is parametrized by its strength B_λ and smoothing length λ, and in accordance with the generation process, it also depends on k_D, k_m, and the spectral index n_B. Now, if this seed is indeed presented during recombination, it prints a signal in the pattern on CMB spectra, signal that depends on the aforementioned variables. Here we have computed the power spectrum in the CMB radiation sourced by this primordial field, and we observed how the shape of this spectrum changes given different values of the magnetic parameters. This means that, by constraining the value of these magnetic parameters via CMB observations, we can have some clue about the mechanism which produced this field. We have also studied the magnetic field at the background level and a first order at the perturbation theory. In the first case, we observed how the presence of PMFs can enhance the speed of sound of the plasma and the time of matter-radiation density equality, producing an increase in the amplitude of the acoustic peaks and shift them to small multipolar numbers. Other important effect of PMFs at the background level is the faster decay of the gravitational potentials when they enter to the sound horizon; this effect could be seen in the subsequent formation of the large-scale structure. Secondly, at first order in the perturbation theory, the scalar mode in the magnetic field produces an increase in the Sachs-Wolfe, although the effect could not be seen observationally due to small effect compared with the primary signal. We also found how the value of the parameters related to B and n_s changes the shape of the power spectrum, and by increasing k_m the peaks related to the ratio between with and without IR cutoff are shifted to large angles. Moreover, in scenarios like inflation, the effect of infrared cutoff might not be ignored (for a deeper discussion see [39]); thus, the feature of this signal will be useful for constraining PMF inflation generation models. In fact, this k_m is important for studying the evolution of density perturbations and peculiar velocities due to primordial magnetic fields and effects on BBN [37, 68, 69]. Additionally, the power spectrum generated by magnetic fields is blue for $n_B > 0$ and red for $n_B < 0$, which means that for causal fields ($n_B > 2$), the signal printed in the CMB spectrum is weak, while for noncausal fields ($n_B < 2$), the signal is more strong, being ($n_B \sim -3$) the maximal value for the field corresponding to a scale invariant spectrum. Therefore, cosmological data are more favorable to noncausal PMFs ($n_B \sim -2.37$) as we can see in the **Figure 6**. In conclusion, the study of PMFs and their effects on different cosmological datasets, mainly in CMB, will provide new insight into the physics at the early Universe and could explain the actual origin of the cosmic magnetic fields.

Conflict of interest

The authors declare that there is no conflict of interest.

Author details

Héctor Javier Hortúa* and Leonardo Castañeda
Universidad Nacional de Colombia, Bogotá, Colombia

*Address all correspondence to: hjhortuao@unal.edu.co

References

[1] Grasso D, Rubinstein HR. Magnetic fields in the early universe. Physics Reports. 2001;**348**(3):163-266

[2] Subramanian K. The origin, evolution and signatures of primordial magnetic fields. Reports on Progress in Physics. 2016;**79**(7):076901

[3] Hortua HJ, Castañeda L. Reduced bispectrum seeded by helical primordial magnetic fields. Journal of Cosmology and Astroparticle Physics. 2017; **2017**(06):020

[4] Mack A, Kahniashvili T, Kosowsky A. Microwave background signatures of a primordial stochastic magnetic field. Physical Review D. 2002;**65**(12):123004

[5] Kahniashvili T, Ratra B. CMB anisotropies due to cosmological magnetosonic waves. Physical Review D. 2007;**75**(2):023002

[6] Hortua HJ, Castañeda L. Parity odd CMB power spectrum via helical magnetic fields. In: Proceedings of the 28th Texas Symposium on Relativistic Astrophysics. 2015

[7] Shiraishi M. Probing the Early Universe with the CMB Scalar, Vector and Tensor Bispectrum. Tokyo: Springer; 2013

[8] Campanelli L. Lorentz-violating inflationary magnetogenesis. The European Physical Journal C. 2015;**75**: 278

[9] Durrer R, Neronov A. Cosmological magnetic fields: Their generation, evolution and observation. The Astronomy and Astrophysics Review. 2013;**21**(1):62

[10] Subramanian K. Magnetic fields in the early universe. Astronomische Nachrichten. 2010;**331**:110

[11] Martin J, Yokoyama J. Generation of large scale magnetic fields in single-field inflation. Journal of Cosmology and Astroparticle Physics. 2008;**1**:025

[12] Ferreira RJZ, Jain R, Sloth MS. Inflationary magnetogenesis without the strong coupling problem. Journal of Cosmology and Astroparticle Physics. 2013;**2013**(10):004

[13] Caprini C, Sorbo L. Adding helicity to inflationary magnetogenesis. Journal of Cosmology and Astroparticle Physics. 2014;**10**:056

[14] Durrer R, Hollenstein L, Jain RK. Can slow roll inflation induce relevant helical magnetic fields? Journal of Cosmology and Astroparticle Physics. 2011;**3**:037

[15] Bamba K, Yokoyama J. Large-scale magnetic fields from inflation in dilaton electromagnetism. Physical Review D. 2004;**69**:043507

[16] Bamba K, Odintsov SD. Inflation and late-time cosmic acceleration in non-minimal Maxwell-F(R) gravity and the generation of large-scale magnetic fields. Journal of Cosmology and Astroparticle Physics. 2008;**4**:024

[17] Adshead P, Giblin JT Jr, Scully TR, Sfakianakis EI. Magnetogenesis from axion inflation. Journal of Cosmology and Astroparticle Physics. 2016;**10**:039

[18] Emami R, Firouzjahi H, Sadegh Movahed M. Inflation from charged scalar and primordial magnetic fields? Physical Review D. 2010;**81**:083526

[19] Ratra B. Cosmological ‘seed’ magnetic field from inflation. The Astrophysical Journal. 1992;**391**:L1-L4

[20] Demozzi V, Mukhanov V, Rubinstein H. Magnetic fields from

inflation? Journal of Cosmology and Astroparticle Physics. 2009;**8**:025

[21] Barnaby N, Namba R, Peloso M. Observable non-gaussianity from gauge field production in slow roll inflation, and a challenging connection with magnetogenesis. Physical Review D. 2012;**85**:123523

[22] Kandus A, Kunze KE, Tsagas CG. Primordial magnetogenesis. Physics Reports. 2011;**505**(1):1-58

[23] Widrow LM. Origin of galactic and extragalactic magnetic fields. Reviews of Modern Physics. 2002;**74**:775-823

[24] Widrow LM, Ryu D, Schleicher DRG, et al. The first magnetic fields. Space Science Reviews. 2012;**166**:37-70

[25] Straumann N. Cosmological Phase Transitions; 2004. ArXiv

[26] Quashnock JM, Loeb A, Spergel DN. Magnetic field generation during the cosmological QCD phase transition. The Astrophysical Journal. 1989;**344**:L49-L51

[27] Cheng B, Olinto AV. Primordial magnetic fields generated in the quark-hadron transition. Physical Review D. 1994;**50**:2421-2424

[28] Tevzadze AG, Kisslinger L, Brandenburg A, Kahniashvili T. Magnetic fields from qcd phase transitions. The Astrophysical Journal. 2012;**759**(1):54

[29] Hogan CJ. Magnetohydrodynamic effects of a first-order cosmological phase transition. Physical Review Letters. 1983;**51**:1488-1491

[30] Harrison ER. Generation of magnetic fields in the radiation era. Monthly Notices of the Royal Astronomical Society. 1970;**147**(3):279

[31] Matarrese S, Mollerach S, Notari A, Riotto A. Large-scale magnetic fields from density perturbations. Physical Review D. 2005;**71**(043502)

[32] Hortua HJ, Castañeda L, Tejeiro JM. Evolution of magnetic fields through cosmological perturbation theory. Physical Review D. 2013;**87**:103531

[33] Nalson E, Christopherson AJ, Malik KA. Effects of non-linearities on magnetic field generation. Journal of Cosmology and Astroparticle Physics. 2014;**1409**:023

[34] Barrow JD, Ferreira PG, Silk J. Constraints on a primordial magnetic field. Physical Review Letters. 1997;**78**: 3610-3613

[35] Ballardini M, Finelli F, Paoletti D. CMB anisotropies generated by a stochastic background of primordial magnetic fields with non-zero helicity. Journal of Cosmology and Astroparticle Physics. 2015;(10):031

[36] Durrer R, Caprini C. Primordial magnetic fields and causality. Journal of Cosmology and Astroparticle Physics. 2003;(11):010

[37] Kim E-J, Olinto AV, Rosner R. Generation of density perturbations by primordial magnetic fields. The Astrophysical Journal. 1996;**468**:28

[38] Yamazaki DG. CMB with the background primordial magnetic field. Physical Review D. 2014;**89**(083528)

[39] Yamazaki DG, Kusakabe M. Effects of power law primordial magnetic field on big bang nucleosynthesis. Physical Review D. 2012;**86**(12):123006

[40] Choudhury S. Inflamagnetogenesis redux: Unzipping sub-Planckian inflation via various cosmoparticle probes. Physics Letters B. 2014;**735**: 138-145

[41] Hortúa HJ, Castañeda L. Power spectrum of post-inflationary primordial magnetic fields. Physical Review D. 2014;**90**:123520

[42] Paoletti D, Finelli F. CMB constraints on a stochastic background of primordial magnetic fields. Physical Review D. 2011;**83**(12):123533

[43] Caprini C, Finelli F, Paoletti D, Riotto A. The cosmic microwave background temperature bispectrum from scalar perturbations induced by primordial magnetic fields. Journal of Cosmology and Astroparticle Physics. 2009;**6**(21)

[44] Trivedi P, Subramanian K, Seshadri TR. Primordial magnetic field limits from cosmic microwave background bispectrum of magnetic passive scalar modes. Physical Review D. 2010;**D82**: 123006

[45] Durrer R, Kunz M. Cosmic microwave background anisotropies from scaling seeds: Generic properties of the correlation functions. Physical Review D. 1998;**57**:R3199-R3203

[46] Brown IA. Primordial magnetic fields in cosmology [PhD thesis]; 2008

[47] Brown I, Crittenden R. Non-gaussianity from cosmic magnetic fields. Physical Review D. 2005;**72**(6):063002

[48] Caprini C, Durrer R, Kahniashvili T. Cosmic microwave background and helical magnetic fields: The tensor mode. Physical Review D. 2004;**69**(6): 063006

[49] Martín-García JM. xAct: Efficient Tensor Computer Algebra for Mathematica; 2002

[50] Pogosian L, Vachaspati T, Winitzki S. Signatures of kinetic and magnetic helicity in the cosmic microwave background radiation. Physical Review D. 2002;**65**(083502)

[51] Paoletti D, Finelli F, Paci F. The scalar, vector and tensor contributions of a stochastic background of magnetic fields to cosmic microwave background anisotropies. Monthly Notices of the Royal Astronomical Society. 2009;**396**: 523-534

[52] Finelli F, Paci F, Paoletti D. Impact of stochastic primordial magnetic fields on the scalar contribution to cosmic microwave background anisotropies. Physical Review D. 2008;**78**(2):023510

[53] Giovannini M, Kunze KE. Magnetized cmb observables: A dedicated numerical approach. Physical Review D. 2008;**77**(063003)

[54] Lewis A, Challinor A, Lasenby A. Efficient computation of cosmic microwave background anisotropies in closed Friedmann-Robertson-Walker models. The Astrophysical Journal. 2000;**538**:473-476

[55] Doran M. CMBEASY: An object oriented code for the cosmic microwave background. Journal of Cosmology and Astroparticle Physics. 2005;(10):011

[56] Kahniashvili T, Kosowsky A, Mack A, Durrer R. CMB signatures of a primordial magnetic field. In: Durrer R, Garcia-Bellido J, Shaposhnikov M, editors. Journal of Cosmology and Astroparticle Physics, Volume 555 of American Institute of Physics Conference Series. 2001. pp. 451-456

[57] Hortúa HJ, Casta Neda L. Effects of primordial magnetic fields on CMB. In: Heavens A, Starck J-L, Krone-Martins A, editors. Statistical Challenges in 21st Century Cosmology, Volume 306 of IAU Symposium. 2014. pp. 159-161

[58] Lesgourgues J. The Cosmic Linear Anisotropy Solving System (CLASS) I: Overview; 2011. ArXiv e-prints

[59] Durrer R. The Cosmic Microwave Background. Cambridge University Press; 2008

[60] Lesgourgues J, Pastor S. Massive neutrinos and cosmology. Physics Reports. 2006;**429**(6):307-379

[61] Audren B, Lesgourgues J, Benabed K, Prunet S. Conservative constraints on early cosmology: An illustration of the Monte Python cosmological parameter inference code. Journal of Cosmology and Astroparticle Physics. 2013;**1302**: 001

[62] Yamazaki DG, Ichiki K, Kajino T, Mathews GJ. Effects of the primordial magnetic field on the CMB. In: American Institute of Physics Conference Series. 2007. pp. 449-452

[63] Kahniashvili T, Lavrelashvili G, Ratra B. CMB temperature anisotropy from broken spatial isotropy due to a homogeneous cosmological magnetic field. Physical Review D. 2008;**78**(6): 063012

[64] Paoletti D. Thesis: Small scale CMB anisotropies with Planck: Constraints on primordial magnetic fields and the impact of foreground/ secondary anisotropy residuals. The International Union of Soil Sciences. 2010;**4**(1):276

[65] Shaw JR, Lewis A. Massive neutrinos and magnetic fields in the early universe. Physical Review D. 2010; **81**(4):043517

[66] Shaw JR, Lewis A. Constraining primordial magnetism. Physical Review D. 2012;**86**(4):043510

[67] Hu W, White M. CMB anisotropies: Total angular momentum method. Physical Review D. 1997;**56**:596-615

[68] Yamazaki DG, Ichiki K, Umezu K-i, Hanayama H. Effect of primordial magnetic field on seeds for large scale structure. Physical Review D. 2006;**74** (123518)

[69] Jedamzik K, Katalinic V, Olinto A. Limit on primordial small-scale magnetic fields from cosmic microwave background distortions. Physical Review Letters. 2000;**85**:700-703

Section 3

Alternatives to the Standard GTR

Chapter 5

Cosmological Solutions to Polynomial Affine Gravity in the Torsion-Free Sector

Oscar Castillo-Felisola, José Perdiguero and Oscar Orellana

Abstract

We find possible cosmological models of the polynomial affine gravity described by connections that are either compatible or not with a metric. When possible, we compare them with those of general relativity. We show that the set of cosmological vacuum solutions in general relativity are a subset of the solutions of polynomial affine gravity. In our model, the cosmological constant appears as an integration constant, and, additionally, we show that some forms of matter can be emulated by the affine structure—even in the metric compatible case. In the case of connections not compatible with a metric, we obtain formal families of solutions, which should be constrained by physical arguments. We show that for a certain parametrisation of the connection, the affine Ricci-flat condition yields the cosmological field equations of general relativity coupled with a perfect fluid, pointing towards a geometrical emulation of—what is interpreted in general relativity as—matter effects.

Keywords: affine gravity, exact solutions, cosmological models

1. Introduction

All of the *fundamental* physics is described by four interactions: electromagnetic, weak, strong and gravitational. The former three are bundled into what is known as *standard model* of particle physics, which explains very accurately the physics at very short scales. These three interactions share common grounds, for example, they are modelled by connections with values in a Lie algebra, they have been successfully quantised and renormalised, and the simplest of them—quantum electrodynamics—gives the most accurate results when compared with the experiments.

On the other hand, the model that explains gravitational interaction (general relativity) is a field theory for the metric, which can be thought as a potential for the gravitational connection [1, 2]. Although general relativity is the most successful theory, we have to explain gravity [3–5], it cannot be formulated as a gauge theory (in four dimensions), the standard quantisation methods lead to inconsistencies, and it is non-renormalisable, driving the community to believe it is an effective theory of a yet unknown fundamental one. Within the framework of cosmology, when one wants to conciliate both *standard models*,[1] it was noticed that nearly 95%

[1] Besides the standard model of particles, there is a *standard model of cosmology*.

of the Universe does not fit into the picture. Therefore, a (huge) piece of the puzzle is missing called the dark sector of the Universe, composed of dark matter and dark energy. In order to solve this problem, one needs to add *new physics*, by either including extra particles (say inspired in beyond standard model physics) or changing the gravitational sector. The latter has inspired plenty of generalisations of general relativity.

Although it cannot be said that the mentioned troubles are due to the fact that the model is described by the metric, given that the *physical* quantity associated with the gravitational interaction—the curvature—is defined for a connection, it is worth to ask ourselves whether a *more* fundamental model of gravitational interactions can be built up using the affine connection as the mediator.

The first affine model of gravity was proposed by Eddington in Ref. [6], where the action was defined by the square root of the determinant of the Ricci tensor:

$$S = \int \sqrt{\det(\mathrm{Ric})}, \tag{1}$$

but in Schrödinger's words [7]:

> *For all that I know, no special solution has yet been found which suggests an application to anything that might interest us...*

However, Eddington's idea serves as a starting point to new proposals [8, 9].

In a series of seminal papers [10–13], Cartan presented a definition of curvature for spaces with torsion and its relevance for general relativity. It is worth mentioning that in pure gravity—described by the Einstein-Hilbert-like action, Cartan's generalisation of gravity yields the condition of vanishing torsion as an equation of motion. Therefore, it was not seriously considered as a generalisation of general relativity, until the inclusion of gravitating fermionic matter [14].

Inspired by Cartan's idea of considering an affine connection into modelling of gravity, a new interesting proposal has been considered. Among the interesting generalisations, we mention a couple: (i) the well-known metric-affine models of gravity [15], in which the metric and connection are not only considered as independent, but the conditions of metricity and vanishing torsion are in general dropped and (ii) the Lovelock-Cartan gravity [16], includes extra terms in the action compatible with the precepts of general relativity, whose variation yields field equations that are second-order differential equations. Nonetheless, the metric plays a very important role in these models.

Modern attempts to describe gravity as a theory for the affine connection have been proposed in Refs. [17–25], and the cosmological implications in an Eddington-inspired affine model were studied in Refs. [26–29].

The recently proposed polynomial affine gravity [24] separates the two roles of the metric field, as in a Palatini formulation of gravity, but does not allow it to participate in the mediation of the interaction, by its exclusion from the action. It turns out that the absence of the metric in the action results in a robust structure that—without the addition of other fields—does not accept deformations. That robustness can be useful if one would like to quantise the theory, because all possible counter-terms should have the form of terms already present in the action.

In this chapter, we focus in finding cosmological solutions in the context of polynomial affine gravity, restricted to torsion-free sector of equi-affine connections, which yields a simple set of field equations generalising those obtained in standard general relativity [25]. This chapter is divided into four sections: In Section 2, we review briefly the polynomial affine model of gravity. In Section 3, we use the

Levì-Civita connection for a Friedman-Robertson-Walker metric, to solve the field equations—obtained in the torsion-free sector—of polynomial affine gravity. Then, in Section 4, we solve for, the case of (affine) Ricci-flat manifold, the field equations for the affine connection. Some remarks and conclusions are presented in Section 5. For completeness, in Appendix A, we include a short exposition of the Lie derivative applied to the connection and show the Killing vectors compatible with the cosmological principle.

2. Polynomial affine gravity

In the standard theory of gravity, general relativity, the fundamental field is the metric, $g_{\mu\nu}$, of the space–time [1, 2]. Nevertheless, the metric has a twofold role in this gravitational model: it measures distances and also defines the notion of parallelism, that is, settles the connection. Palatini, in Ref. [30], considered a somehow separation of these roles, but at the end of the day, the metric was still the sole field of the model. It was understood soon after that the connection, $\Gamma^{\mu}{}_{\rho\sigma}$, does not need to be related with the metric field [10–13, 31], and, therefore, the curvature could be blind to the metric.

In this section, we briefly expose the model proposed in Refs. [24, 25], which is inspired in the aforementioned role separation. The metric is left out the mediation of gravitational interactions by taking it out the action.

The action of the polynomial affine gravity is built up from an affine connection, $\hat{\Gamma}^{\mu}{}_{\rho\sigma}$, which accepts a decomposition on irreducible components as

$$\hat{\Gamma}^{\mu}{}_{\rho\sigma} = \hat{\Gamma}^{\mu}{}_{(\rho\sigma)} + \hat{\Gamma}^{\mu}{}_{[\rho\sigma]} = \Gamma^{\mu}{}_{\rho\sigma} + \varepsilon_{\rho\sigma\lambda\kappa} T^{\mu,\,\lambda\kappa} + A_{[\rho}\delta^{\mu}_{\nu]}, \tag{2}$$

where $\Gamma^{\mu}{}_{\rho\sigma} = \hat{\Gamma}^{\mu}{}_{(\rho\sigma)}$ is symmetric in the lower indices, A_{ρ} is a vector field corresponding to the trace of torsion, and $T^{\mu,\,\lambda\kappa}$ is a Curtright field [32], which satisfy the properties $T^{\kappa,\,\mu\nu} = -T^{\kappa,\,\nu\mu}$ and $\varepsilon_{\lambda\kappa\mu\nu}T^{\kappa,\,\mu\nu} = 0$. The metric field, which might or might not exist, cannot be used for contracting nor lowering or raising indices. The; relation between the epsilons with lower and upper indices is given by $\epsilon^{\delta\eta\lambda\kappa}\epsilon_{\mu\nu\rho\sigma} = 4!\delta^{\delta}{}_{[\mu}\delta^{\eta}{}_{\nu}\delta^{\lambda}{}_{\rho}\delta^{\kappa}{}_{\sigma]}$.

The most general action preserving diffeomorphism invariance, written in terms of the fields in Eq. (2), is

$$\begin{aligned} S[\Gamma,T,A] = \int \mathrm{d}^4x \Big[& B_1 R_{\mu\nu}{}^{\mu}{}_{\rho} T^{\nu,\,\alpha\beta} T^{\rho,\,\gamma\delta} \epsilon_{\alpha\beta\gamma\delta} + B_2 R_{\mu\nu}{}^{\sigma}{}_{\rho} T^{\beta,\,\mu\nu} T^{\rho,\,\gamma\delta} \epsilon_{\sigma\beta\gamma\delta} + B_3 R_{\mu\nu}{}^{\mu}{}_{\rho} T^{\nu,\,\rho\sigma} A_{\sigma} + B_4 R_{\mu\nu}{}^{\sigma}{}_{\rho} T^{\rho,\,\mu\nu} A_{\sigma} \\ & + B_5 R_{\mu\nu}{}^{\rho}{}_{\rho} T^{\sigma,\,\mu\nu} A_{\sigma} + C_1 R_{\mu\rho}{}^{\mu}{}_{\nu} \nabla_{\sigma} T^{\nu,\,\rho\sigma} + C_2 R_{\mu\nu}{}^{\rho}{}_{\rho} \nabla_{\sigma} T^{\sigma,\,\mu\nu} + D_1 T^{\alpha,\,\mu\nu} T^{\beta,\,\rho\sigma} \nabla_{\gamma} T^{(\lambda,\kappa)\gamma} \epsilon_{\beta\mu\nu\lambda} \epsilon_{\alpha\rho\sigma\kappa} \\ & + D_2 T^{\alpha,\,\mu\nu} T^{\lambda,\,\beta\gamma} \nabla_{\lambda} T^{\delta,\,\rho\sigma} \epsilon_{\alpha\beta\gamma\delta} \epsilon_{\mu\nu\rho\sigma} + D_3 T^{\mu,\,\alpha\beta} T^{\lambda,\,\nu\gamma} \nabla_{\lambda} T^{\delta,\,\rho\sigma} \epsilon_{\alpha\beta\gamma\delta} \epsilon_{\mu\nu\rho\sigma} + D_4 T^{\lambda,\,\mu\nu} T^{\kappa,\,\rho\sigma} \nabla_{(\lambda} A_{\kappa)} \epsilon_{\mu\nu\rho\sigma} \\ & + D_5 T^{\lambda,\,\mu\nu} \nabla_{[\lambda} T^{\kappa,\,\rho\sigma} A_{\kappa]} \epsilon_{\mu\nu\rho\sigma} + D_6 T^{\lambda,\,\mu\nu} A_{\nu} \nabla_{(\lambda} A_{\mu)} + D_7 T^{\lambda,\,\mu\nu} A_{\lambda} \nabla_{[\mu} A_{\nu]} \\ & + E_1 \nabla_{(\rho} T^{\rho,\,\mu\nu} \nabla_{\sigma)} T^{\sigma,\,\lambda\kappa} \epsilon_{\mu\nu\lambda\kappa} + E_2 \nabla_{(\lambda} T^{\lambda,\,\mu\nu} \nabla_{\mu)} A_{\nu} + T^{\alpha,\,\beta\gamma} T^{\delta,\,\eta\kappa} T^{\lambda,\,\mu\nu} T^{\rho,\,\sigma\tau} \big(F_1 \epsilon_{\beta\gamma\eta\kappa} \epsilon_{\alpha\rho\mu\nu} \epsilon_{\delta\lambda\sigma\tau} \\ & + F_2 \epsilon_{\beta\lambda\eta\kappa} \epsilon_{\gamma\rho\mu\nu} \epsilon_{\alpha\delta\sigma\tau} \big) + F_3 T^{\rho,\,\alpha\beta} T^{\gamma,\,\mu\nu} T^{\lambda,\,\sigma\tau} A_{\tau} \epsilon_{\alpha\beta\gamma\lambda} \epsilon_{\mu\nu\rho\sigma} + F_4 T^{\eta,\,\alpha\beta} T^{\kappa,\,\gamma\delta} A_{\eta} A_{\kappa} \epsilon_{\alpha\beta\gamma\delta} \Big], \end{aligned} \tag{3}$$

where terms related through partial integration and topological invariant have been dropped.[2] One can prove via a dimensional analysis, the *uniqueness* of the above action (see Ref. [25]).

The action in Eq. (3) shows up very interesting features: (i) it is power-counting renormalisable;[3] (ii) all coupling constants are dimensionless which hints the

---

[2] An example of four-dimensional topological term is the Euler density.

[3] Power-counting renormalisability does not guarantee renormalisability.

conformal invariance of the model [33]; (iii) yields no three-point graviton vertices, which might allow to overcome the *no-go* theorems found in Refs. [34, 35]; (iv) its non-relativistic geodesic deviation agrees with that produced by a Keplerian potential [24]; and (v) the effective equations of motion in the torsion-free limit are a generalisation of Einstein's equations [25]. In the remaining of this section, we will sketch how to find the relativistic limit of this model, when the torsion vanishes.

First, notice that the vanishing torsion condition is equivalent to setting both $T^{\lambda,\mu\nu}$ and A_μ equal to zero. Although this limit is not well defined at the action level, it is well defined at the level of equation of motion.[4] In order to simplify the task of finding the equations of motion to take the limit, we restrict ourselves to the terms in the action which are linear in either $T^{\lambda,\mu\nu}$ or A_μ, since these are the only terms which, after the extremisation, will survive the torsion-free limit. Therefore, after the described considerations, the effective torsion-free action is

$$S_{\text{eff}} = \int \mathrm{d}^4x \left(C_1 R_{\lambda\mu}{}^{\lambda}{}_{\nu}\nabla_\rho + C_2 R_{\mu\rho}{}^{\lambda}{}_{\lambda}\nabla_\nu\right)T^{\nu,\mu\rho}. \tag{4}$$

The nontrivial equations of motion for this action are those for the Curtright field, $T^{\nu,\mu\rho}$:

$$\nabla_{[\rho}R_{\mu]\nu} + \kappa\nabla_\nu R_{\mu\rho}{}^{\lambda}{}_{\lambda} = 0, \tag{5}$$

where κ is a constant related with the original couplings of the model.

In the Riemannian formulation of differential geometry, since the curvature tensor is antisymmetric in the last couple of indices, the second term in Eq. (5) vanishes identically. However, for non-Riemannian connections, such term still vanishes if the connection is compatible with a volume form. These connections are known as equi-affine connections [36, 37]. In addition, the Ricci tensor for equi-affine connections is symmetric. For these connections, the gravitational equations are simply

$$\nabla_{[\rho}R_{\mu]\nu} = 0. \tag{6}$$

Eq. (6) is a generalisation of the parallel Ricci curvature condition, $\nabla_\rho R_{\mu\nu} = 0$, which is a known extension of Einstein's equations [38, 39]. Moreover, these field equations are also obtained as part of a Ã la Palatini approach to a Yang-Mills formulation of gravity, known as the Stephenson-Kilmister-Yang (or SKY) model, proposed in Refs. [40–42]. Such Yang-Mills-like gravity is described by the action

$$S_{sky} = \int \mathrm{d}^4x\sqrt{g}g^{\mu\nu}g^{\sigma\tau}R_{\mu\sigma}{}^{\lambda}{}_{\rho}R_{\nu\tau}{}^{\rho}{}_{\lambda}, \tag{7}$$

which can be written using the curvature two-form as

$$S_{sky} = \int \mathrm{Tr}(\boldsymbol{\mathcal{R}}\star\boldsymbol{\mathcal{R}}) = \int (\boldsymbol{\mathcal{R}}^a_b\star\boldsymbol{\mathcal{R}}^b_a). \tag{8}$$

Although the field equations of the connection obtained from Eq. (7) are the harmonic curvature condition [43],

---

[4] The field equations can be consistently truncated under the requirement of vanishing torsion. It is worth noticing that this condition does not yield the Riemannian theory, since we are not yet asking for a metricity condition.

$$\nabla_\lambda R_{\mu\nu}{}^{\lambda}{}_{\rho} = 0, \tag{9}$$

these are equivalent to Eq. (6) through the second Bianchi identity [39, 44].

The Stephenson-Kilmister-Yang model is a field theory for the metric—not for the connection, and thus there is an extra field equation for the metric. The field equation for the metric is very restrictive, and it does not accept Schwarzschild-like solutions [45]. However, in the polynomial affine gravity, since the metric does not participate in the mediation of gravitational interaction, that problem is solved trivially. Meanwhile, the physical field associated with the gravitational interaction is the connection. This difference makes a huge distinction in the phenomenological interpretation of these models.

In the following sections, we shall present solutions to the field Eqs. (6), in the cases where the connection is metric or not. To this end, in Appendix A we show how to propose an ansatz compatible with the desired symmetries. Moreover, Eq. (6) can be solved in three ways, yielding to a sub-classification of the solutions: (i) Ricci flat solutions, $R_{\mu\nu} = 0$; (ii) parallel Ricci solutions, $\nabla_\lambda R_{\mu\nu} = 0$; and (iii) harmonic Riemann solutions, $\nabla_\lambda R_{\mu\nu}{}^{\lambda}{}_{\rho} = 0$.

3. Cosmological metric solutions

The conditions of isotropy and homogeneity are very stringent, when imposed on a symmetric rank-two tensor, and the possible ansatz is just the Friedmann-Robertson-Walker metric:

$$g = G_{00}(t)\mathrm{d}t\otimes\mathrm{d}t + G_{11}(t)\left(\frac{1}{1-\kappa r^2}\mathrm{d}r\otimes\mathrm{d}r + r^2\mathrm{d}\theta\otimes\mathrm{d}\theta + r^2\sin(\theta)^2\mathrm{d}\phi\otimes\mathrm{d}\phi\right). \tag{10}$$

In the remaining of this section, we shall use the standard parametrisation of a Friedmann-Robertson-Walker metric, that is

$$g = -\mathrm{d}t\otimes\mathrm{d}t + a^2(t)\left(\frac{1}{1-\kappa r^2}\mathrm{d}r\otimes\mathrm{d}r + r^2\mathrm{d}\theta\otimes\mathrm{d}\theta + r^2\sin(\theta)^2\mathrm{d}\phi\otimes\mathrm{d}\phi\right). \tag{11}$$

The nonvanishing components of the Levi-Cività connection for the metric in Eq. (11) are

$$\begin{array}{lll}
\Gamma^{t}{}_{rr} = -\dfrac{a\dot a}{kr^2-1} & \Gamma^{t}{}_{\theta\theta} = r^2 a\dot a & \Gamma^{t}{}_{\phi\phi} = r^2 a\sin^2(\theta)\dot a \\
\Gamma^{r}{}_{tr} = \dfrac{\dot a}{a} & \Gamma^{r}{}_{rt} = \dfrac{\dot a}{a} & \Gamma^{r}{}_{rr} = -\dfrac{kr}{kr^2-1} \\
\Gamma^{r}{}_{\theta\theta} = kr^3 - r & \Gamma^{r}{}_{\phi\phi} = \left(kr^3-r\right)\sin^2(\theta) & \Gamma^{\theta}{}_{t\theta} = \dfrac{\dot a}{a} \\
\Gamma^{\theta}{}_{r\theta} = \dfrac{1}{r} & \Gamma^{\theta}{}_{\theta t} = \dfrac{\dot a}{a} & \Gamma^{\theta}{}_{\theta r} = \dfrac{1}{r} \\
\Gamma^{\theta}{}_{\phi\phi} = -\cos(\theta)\sin(\theta) & \Gamma^{\phi}{}_{t\phi} = \dfrac{\dot a}{a} & \Gamma^{\phi}{}_{r\phi} = \dfrac{1}{r} \\
\Gamma^{\phi}{}_{\theta\phi} = \dfrac{\cos(\theta)}{\sin(\theta)} & \Gamma^{\phi}{}_{\phi t} = \dfrac{\dot a}{a} & \Gamma^{\phi}{}_{\phi r} = \dfrac{1}{r} \\
 & \Gamma^{\phi}{}_{\phi\theta} = \dfrac{\cos(\theta)}{\sin(\theta)} &
\end{array} \tag{12}$$

3.1 … with vanishing Ricci

This particular case is a metric model of gravity, whose field equations are vanishing Ricci. It is expected to obtain the cosmological vacuum solution of general relativity (without cosmological constant), that is, Minkowski space–time.

From the connection in Eq. (12), it is straightforward to calculate the Ricci tensor, and the field equations are then

$$R_{tt} = -\frac{3\ddot{a}}{a} = 0, \qquad R_{ii} = f_i(r,\theta)\left(\dot{a}^2 + a\ddot{a} + 2\kappa\right) = 0, \tag{13}$$

where the functions f_i are $f_r = \left(1 - \kappa r^2\right)^{-1}$, $f_\theta = r^2$ and $f_\varphi = r^2 \sin^2(\theta)$.

The solutions to Eq. (13) are shown in **Table 1** and (as expected) are two parametrisations of Minkowski space–time (see, for example, Ref. [46]).

3.2 … with parallel Ricci

Secondly, we shall analyse the possible solutions to the parallel Ricci equations:

$$\nabla_\lambda R_{\mu\nu} = 0. \tag{14}$$

Notice that in the case of Riemannian geometry, there is a *natural* parallel symmetric $\begin{pmatrix} 0 \\ 2 \end{pmatrix}$-type tensor, that is, the metric. Therefore, a simple solution to Eq. (14) is that the Ricci is proportional to the metric—the space–time is an Einstein manifold, and the proportionality factor is related with the cosmological constant.

The independent components of Eq. (14) for the ansätze in Eq. (11) are

$$\nabla_t R_{tt} \simeq \dot{a}\ddot{a} - a\dddot{a} = 0, \tag{15}$$

$$\nabla_i R_{ti} \simeq \left(\dot{a}^2 - a\ddot{a} + \kappa\right)\dot{a} = 0. \tag{16}$$

Additionally, Eq. (15) can be rewritten as

$$\frac{d}{dt}\left(\frac{\ddot{a}}{a}\right) = 0 \quad \Rightarrow \quad \ddot{a} + Ca = 0. \tag{17}$$

According to the value of the integration constant C, we parametrise it as

$$C = \begin{cases} \omega^2 & \text{for } C > 0 \\ \omega = 0 & \text{for } C = 0 \\ v & \text{for } C < 0 \end{cases}$$

Using Eq. (17) to eliminate the $\ddot{a}$ dependence from Eq. (16) yields

$$\dot{a}^2 + Ca^2 + \kappa = 0. \tag{18}$$

Scale factor for the metric vanishing Ricci case		
$\kappa = -1$	$\kappa = 0$	$\kappa = 1$
$\sqrt{2}t + B$	$B \in \mathbb{R}^+$	$\nexists$

Table 1.
Scale factor solving the vanishing Ricci condition, for a cosmological metric connection.

The solutions to Eq. (17) are presented in **Table 2**, and they are known from general relativity (see, for example, Ref. [46]). Interestingly, our integration constant, C, could be identified as $C = -\frac{\Lambda}{3}$ from the vacuum Friedmann's equations. However, our equations are compatible with Friedmann's equations, interacting with a vacuum energy perfect fluid, if the integration constant is identified with

$$C = \frac{4\pi G_N}{3}(\rho + 3p) - \frac{\Lambda}{3}. \tag{19}$$

3.3 ... with harmonic Riemann

Now that we showed that the solutions of the parallel Ricci equations are equivalent to those of general relativity, we turn our attention to Eq. (6). For the metric ansatz in Eq. (11), interestingly, only an independent equation is obtained:

$$\frac{2\dot{a}^3 - a\dot{a}\ddot{a} - a^2\dddot{a} + 2\kappa\dot{a}}{a} = 0,$$

that should determine the scale factor. It can be rewritten as

$$-\frac{d}{dt}\left(\frac{\ddot{a}}{a} + \frac{\dot{a}^2}{a^2} + \frac{\kappa}{a^2}\right) = 0, \tag{20}$$

that is,

$$\frac{\ddot{a}}{a} + \frac{\dot{a}^2}{a^2} + \frac{\kappa}{a^2} = -C. \tag{21}$$

After a change of variable, $f = a^2$, Eq. (21) becomes

$$\ddot{f} + 2Cf + 2\kappa = 0, \tag{22}$$

whose solutions are

$$f(t) = \begin{cases} -\kappa t^2 + At + B & C = 0 \\ A\sin(\omega t) + B\cos(\omega t) - \dfrac{2\kappa}{\omega^2} & 2C = \omega^2 > 0 \\ A\sinh(\omega t) + B\cosh(\omega t) + \dfrac{2\kappa}{\omega^2} & 2C = -\omega^2 < 0 \end{cases} \tag{23}$$

Therefore, the scale factors are those presented in **Table 3**. Notice, however, that in this case we are not separating the cases according to the value of κ, but the

Scale factor for the metric parallel Ricci case			
	$\kappa = -1$	$\kappa = 0$	$\kappa = 1$
$C = -\omega^2 < 0$	$\pm\frac{\sinh(\omega t)}{2\omega}$	$A\exp(\pm\omega t)$	$\pm\frac{\cosh(\omega t)}{2\omega}$
$C = \omega = 0$	$\pm t + B$	$B \in^{+}$	$\nexists$
$C = \omega^2 > 0$	$\frac{\sin(\omega t + \varphi)}{\omega}$	$\nexists$	$\nexists$

Table 2.
Scale factor solving the parallel Ricci condition, for a cosmological metric connection.

Scale factor for the metric harmonic curvature case	
	$\kappa = -1, 0, 1$
$C = -\omega^2 < 0$	$\sqrt{A\sinh(\omega t) + B\cosh(\omega t) + \frac{2\kappa}{\omega^2}}$
$C = \omega = 0$	$\sqrt{-\kappa t^2 + At + B}$
$C = \omega^2 > 0$	$\sqrt{\frac{\sin(\omega t + \varphi)}{\omega}}$

Table 3.
Scale factor solving the harmonic curvature condition, for a cosmological metric connection.

existence of a solution for a given κ is determined by the domain of time and also by the values of the integration constants A and B.

4. Cosmological nonmetric solutions

In order to solve the set of coupled, non-linear, partial differential equations for the connection, one proceeds—just as in general relativity—by giving an ansatz compatible with the symmetries of the problem. Using the Lie derivative, we have found the most general torsion-free connection compatible with the cosmological principle [47].[5] The nonvanishing components of the connection are

$$\begin{array}{lll}
\Gamma^t{}_{tt} = f(t) & \Gamma^t{}_{rr} = \dfrac{g(t)}{1-\kappa r^2} & \Gamma^t{}_{\theta\theta} = r^2 g(t) \\
\Gamma^t{}_{\phi\phi} = r^2 g(t)\sin^2(\theta) & \Gamma^r{}_{tr} = h(t) & \Gamma^r{}_{rt} = h(t) \\
\Gamma^r{}_{rr} = \dfrac{\kappa r}{1-\kappa r^2} & \Gamma^r{}_{\theta\theta} = \kappa r^3 - r & \Gamma^r{}_{\phi\phi} = (\kappa r^3 - r)\sin^2(\theta) \\
\Gamma^\theta{}_{t\theta} = h(t) & \Gamma^\theta{}_{r\theta} = \dfrac{1}{r} & \Gamma^\theta{}_{\theta t} = h(t) \\
\Gamma^\theta{}_{\theta r} = \dfrac{1}{r} & \Gamma^\theta{}_{\phi\phi} = -\cos(\theta)\sin(\theta) & \Gamma^\phi{}_{t\phi} = h(t) \\
\Gamma^\phi{}_{r\phi} = \dfrac{1}{r} & \Gamma^\phi{}_{\theta\phi} = \dfrac{\cos(\theta)}{\sin(\theta)} & \Gamma^\phi{}_{\phi t} = h(t) \\
\Gamma^\phi{}_{\phi r} = \dfrac{1}{r} & \Gamma^\phi{}_{\phi\theta} = \dfrac{\cos(\theta)}{\sin(\theta)} &
\end{array} \tag{24}$$

with f, g and h the unknown functions of time to be determined. The Levi-Cività connection compatible with the Friedman-Robertson-Walker metric is obtained from Eq. (24) by setting $f = 0, g = a\dot{a}$ and $h = \frac{\dot{a}}{a}$—compare with Eq. (12).

The Ricci tensor calculated for the connection in Eq. (24) has only two independent components:

$$R_{tt} = 3fh - 3h^2 - 3\frac{\partial h}{\partial t}, \tag{25}$$

$$R_{ii} \simeq fg + gh + 2\kappa + \frac{\partial g}{\partial t}. \tag{26}$$

[5] See Appendix A for a brief comment about the Lie derivative of a connection.

We now proceed to find solutions to Eq. (6). As in the previous section, we present the three possibilities of solutions, but we will restrict ourselves to finding solutions to the (affine) Ricci-flat case.

4.1 ... with vanishing Ricci

The first kind of solutions can be found by solving the system of equations determined by vanishing Ricci. However, this strategy requires the fixing of one of the unknown functions. The equations to solve are written as

$$\dot{h} - (f - h)h = 0, \tag{27}$$

$$\dot{g} + (f + h)g + 2\kappa = 0. \tag{28}$$

Noticing that in the above equations f is not a dynamical function, from Eq. (27) we can solve h as a function of f:

$$h(t) = \frac{\exp(F(t))}{C_h + \int dt \exp(F)}, \tag{29}$$

where we have defined $F = \int dt f(t)$ and C_h is an integration constant. Then, Eq. (28) can be solved for g:

$$g(t) = \exp(-\Sigma(t))\left(C_g - 2\kappa\left(\int dt \exp(\Sigma(t))\right)\right), \tag{30}$$

where $\Sigma(t) = \int dt (f(t) + h(t))$ and C_g is another integration constant.

A particular solution inspired in the components of the connection for Friedmann-Robertson-Walker, in whose case $f = 0$, gives

$$f(t) = 0,\ \ g(t) = \frac{1}{t + C_h}\left(C_g - \kappa(t + C_h)^2\right),\ \ h(t) = \frac{1}{t + C_h}, \tag{31}$$

which for $C_h = C_g = 0$ and $\kappa = -1$ yields the expected solution from **Table 1.**[6] However, in Eq. (30) there are Ricci-flat solutions which cannot be associated with the sole existence of a metric, that is, non-Riemannian manifolds, as, for example, solutions with $\kappa > 0$.

There are special solutions that cannot be obtained from Eqs. (29) and (30), since they represent degenerated point in the moduli space.

Case $f = h$: In these particular subspaces on the moduli, the first equation is linear, and therefore the solution above is not valid. However, the solutions to Eqs. (27) and (28) are given by

$$f = C_h \qquad h = C_h \qquad g = C_g \exp(-2Ct) - \frac{\kappa}{C_h}. \tag{32}$$

Case $h = -f$: In this case again, Eq. (27) decouples from Eq. (28), and there solutions are given by

[6] The standard parametrisation of Minkowski space–time is achieved by the trivial solution of Eqs. (27) and (28), i.e. $f = g = h = \kappa = 0$.

$$f=-\frac{1}{2t+C_h}\qquad h=\frac{1}{2t+C_h}\qquad g=C_g-2\kappa t \tag{33}$$

Case $h=0$ and f given: In this case, Eq. (28) becomes an identity, and g can still be solved for a given function f as

$$g(t)=\exp\left(-F(t)\right)\left(C_g-2\kappa\left(\int dt\exp\left(F(t)\right)\right)\right). \tag{34}$$

Case $g=0$ and f given: In this case, Eq. (28) requires $\kappa=0$, and h can still be solved for a given function f as in Eq. (29).

At this point, we have shown that a space–time described by a Ricci flat, torsion-free, equi-affine connection with the form presented in Eq. (24) reproduces the cosmological Ricci-flat solutions to general relativity, presented in **Table 1**, and there exist generalisations to these solutions which are not possibly obtained in the Riemannian case. However, one can go even further and ask oneself whether the affine Ricci-flat condition yields more—real life—useful solutions, such as those solutions of general relativity presented in **Table 2**.

Therefore, we would like to obtain the Einstein equations from the affine Ricci-flat equation, that is,

$$R^{\text{Aff}}_{\mu\nu}=M^{\text{GR}}_{\mu\nu}=0, \tag{35}$$

where

$$M^{\text{GR}}_{\mu\nu}=R^{\text{GR}}_{\mu\nu}-\Lambda g_{\mu\nu}-8\pi G\left(T_{\mu\nu}-\frac{1}{2}Tg_{\mu\nu}\right).$$

In the following, we are considering that the stress-energy tensor describes a perfect fluid, that is,

$$T_{\mu\nu}=\text{diag}\left(\rho\quad\frac{pa^2}{1-\kappa r^2}\quad pa^2r^2\quad pa^2r^2\sin^2\theta\right).$$

In general relativity, the Einstein equations in the form of Ricci, for the cosmological ansatz, yield

$$M^{\text{GR}}_{tt}\simeq 3\ddot{a}-\Lambda a+4\pi Ga(\rho+3p),\qquad M^{\text{GR}}_{tt}\simeq a\ddot{a}+2\dot{a}^2-\Lambda a^2-4\pi Ga^2(\rho-p)+2\kappa, \tag{36}$$

Now, comparing Eq. (36) with Eqs. (27) and (28), a parametrisation for f,g and h can be found such that once one computes the Ricci tensor for the affine connection, the compatibility in Eq. (35) is satisfied. The parametrisation is given by

$$h=\dot{a}+x,\qquad f=x,\qquad g=a\dot{a}+y, \tag{37}$$

where the functions x and y satisfy the equations

$$\dot{x}+x\dot{a}-F_1=0, \tag{38}$$

$$\dot{y}+yF_2-F_3=0, \tag{39}$$

with

$$F_1=\frac{a}{3}(4\pi G(3p+\rho)-\Lambda)-(\dot{a})^2,\qquad F_2=\dot{a}+2x,\qquad F_3=a^2(4\pi G(p-\rho)-\Lambda)-2a\dot{a}x-a(\dot{a})^2+(\dot{a})^2. \tag{40}$$

Eqs. (38) and (39) can be formally integrated in terms of functions a,ρ and p, yielding

$$x = e^{-a}\left(C_x + \int dtF_1 e^a\right), \tag{41}$$

$$y = e^{-\int dtF_2}\left(C_y + \int dtF_3 e^{\int dtF_2}\right). \tag{42}$$

Therefore, a subspace of the possible solutions of the affine Ricci-flat geometries describes the cosmological scenarios from general relativity coupled with perfect fluids. However, the explicit expressions for Eqs. (41) and (42) for obtaining specific solutions to Friedmann-Lemaître-Robertson-Walker models are very complicated.

4.2 ... with parallel Ricci

The second class of solutions can be found by solving the parallel Ricci equation, $\nabla_\lambda R_{\mu\nu} = 0$, which yield three independent field equations:

$$\nabla_t R_{tt} \simeq \ddot{h} - (3f - 2h)\dot{h} - h\dot{f} + 2fh(f - h), \tag{43}$$

$$\nabla_i R_{ti} = \nabla_i R_{it} \simeq 3g\dot{h} - h\dot{g} - 2gh(2f - h) - 2\kappa h, \tag{44}$$

$$\nabla_t R_{ii} \simeq \ddot{g} + g\left(\dot{f} + \dot{h}\right) + (f - h)\dot{g} - gh(f + h) - 4\kappa h. \tag{45}$$

However, the system of equations is complicated enough to avoid an analytic solution.

Despite the complication, we can try a couple of assumptions that simplify the system of equations, for example, if one considers the parametrisation inspired in the Friedmann-Robertson-Walker results, that is, setting $f = 0$, and can solve h from Eq. (43), which is a total derivative in this particular case. Nonetheless, despite the value of the first integration constant, the system of equations imposes that both κ and g vanish.

4.3 ... with harmonic curvature

Finally, the third class of solutions are those of Eq. (6). The set of equations degenerate and yield a single independent field equation:

$$\nabla_\lambda R_{ti}{}^{\lambda}{}_{i} \simeq \ddot{g} + g\dot{f} + f\dot{g} - 2g\dot{h} + 2gh(f - 2h) - 2\kappa h \tag{46}$$

Therefore, we need to set two out of the three unknown functions to be able to solve for the connection.

5. Conclusions and remarks

In this chapter, we have shortly reviewed the recently proposed model of polynomial affine gravity, which is an alternative model for gravitational interactions described solely by the connection, that is, the metric does not play any role in the mediation of the interaction. Among the features of the model, one encounters that despite the numerous possible terms in the action (see Eq. (3)), the absence of a metric tensor gives a sort of rigidity to the action, in the sense that only a very restricted set of terms can be added. Such rigidity suggests that if one attempts to quantise the model, it could be renormalisable. Additionally, all of the coupling constants, in the pure gravity regime, are dimensionless, pointing to a possible conformal invariance of the (pure) gravitational interaction.[7]

Restricting ourselves to equi-affine, torsion-free connections, the field equations are a generalisation of those from general relativity (Eq. (6)). We solved the field equations for an isotropic and homogeneous connection, either compatible with a metric or not. These solutions are classified under three conditions: Ricci flat, parallel Ricci and harmonic curvature.

When the affine connection is the Levi-Cività connection for a Friedman-Robertson-Walker metric, we show that the sole solution for a Ricci-flat space–time is described by the connection of Minkowski's space (see **Table 1**). In the parallel Ricci case, we show that—as intuitively expected—one recovers the vacuum cosmological models from general relativity (see **Table 2**), where the cosmological constant enters as an integration constant, but such constant could be interpreted as (partially) coming from the stress-energy tensor of a vacuum energy perfect fluid, as mentioned—in the context of general relativity—in Ref. [48]. Finally, the (formal) solutions to the harmonic curvature are presented in **Table 3**, but yet some work remains to be done to extract the phenomenology from these solutions.

In the case of the cosmological affine connection, we found that the Ricci-flat condition yields only two independent equations, which are not enough to find the three unknown functions that parametrise the homogeneous and isotropic connection. Nonetheless, since f is not a dynamical function, it serves as a *parametric* function to solve the remaining two, that is, g and h. Interestingly, the three functions can be chosen in a way that Ricci-flat condition for the affine connection yields the Friedmann-Lemaître equations from general relativity coupled with a perfect fluid. In this sense, the pure polynomial affine gravity supersizes general relativity, since geometrically it can mimic effects that are usually interpreted as *matter* effects. However, among the possible solutions for the Ricci-flat condition, there are countless (yet) nonphysical solutions, and what is more, there is nothing that favours the specific choice in Eq. (37) over others. Such landscape drives us to think that another type of condition should be used to restrict even further the possible solutions for the affine connection.

The conditions of affine parallel Ricci could be the cornerstone in solving the aforementioned degeneracy, since these conditions raise three independent equations that would serve to determine the three unknown functions. However, at the moment we have not yet achieved any interesting result in pursuing this goal.

On the other hand, the harmonic curvature condition yields a sole (independent) field equation, and therefore the solutions are even more degenerated than those from the Ricci-flat condition, leaving even more space for nonphysical solutions.

Our research has stressed the importance of considering the connection as the mediator of the gravitational interactions. We have confirmed that in the framework of polynomial affine gravity, the cosmological solutions described by a connection compatible with a Friedmann-Robertson-Walker metric are compatible with those of general relativity, with the possible exception of the case of harmonic curvature. The impact of our contribution lies in the fact that for a generic affine connection, even the simplest condition—Ricci flatness—allows solutions which are (dynamically) equivalent to the system of Friedmann-Lamaître-Robertson-Walker for a perfect fluid (in general relativity), despite the absence of matter in the affine model.

We would like to finish our discussion highlighting that the geometric emulation of matter content can serve as a starting point to a change of paradigm related with the interpretation of the matter content of the Universe, in particular the *dark*

[7] At least at classical level.

sector. We think that our findings might be useful for providing a gravitational origin interpretation of the dark matter and/or energy, driven by the inclusion of extra degrees of freedom coming from the nonmetricity of the connection. Further studies, which take the observations reported in Refs. [49, 50] into account, will need to be performed, to be able of discern between the possibilities of, for example, dark matter that has been originated as a gravitational versus matter effect. Similar analysis should be carried with the dark energy [51, 52].

Acknowledgements

We would like to thank the following people for their support; without those help this work would never have been possible: Aureliano Skirzewski, Cristóbal Corral, Iván Schmidt, Alfonso R. Zerwekh, Claudio Dib, Stefano Vignolo and Jorge Zanelli. We are particularly grateful to the developers of the software SageMath [53], SageManifolds [54–56] and Cadabra2 [57–59], which were used extensively along the development of this work. We gratefully acknowledge the constructive comments of the anonymous referee. The "Centro Científico y Tecnológico de Valparaíso" (CCTVal) is funded by the Chilean Government through the Centers of Excellence Base Financing Program of CONICYT, under project No. FS0821.

A. Lie derivative and killing vectors

The usual procedure for solving Einstein's equation is to propose an ansatz for the metric. That ansatz must be compatible with the symmetries we would like to respect in the problem. A first application is seen in Schwarzschild's metric [60], which is the *most general* symmetric rank-two tensor compatible with the rotation group in three dimensions, an thus is *spherically symmetric*.

The formal study of the symmetries of the fields is accomplished via the Lie derivative (for reviews, see Refs. [61–64]). Below, we briefly explain the use of the Lie derivative for obtaining ansatzes for either the metric or the connection.

The Lie derivative of a connection possesses an inhomogeneous part, in comparison with the one of a rank-three tensor. This can be written schematically as

$$\mathcal{L}_\xi \Gamma^a{}_{bc} = \mathcal{L}_\xi T^a{}_{bc} + \frac{\partial^2 \xi^a}{\partial x^b \partial x^c},$$

or explicitly

$$\mathcal{L}_\xi \Gamma^a{}_{bc} = \xi^m \partial_m \Gamma^a{}_{bc} - \Gamma^m{}_{bc} \partial_m \xi^a + \Gamma^a{}_{mc} \partial_b \xi^m + \Gamma^a{}_{bm} \partial_c \xi^m + \frac{\partial^2 \xi^a}{\partial x^b \partial x^c}, \tag{47}$$

where ξ is the vector defining the symmetry flow.

In particular, for cosmological applications, one asks for *isotropy* and *homogeneity*, which in four dimensions restricts the isometry group to either $SO(4)$, $SO(3,1)$ or $ISO(3)$. The algebra of these groups can be obtained from the algebra $\mathfrak{so}(4)$ through a $3+1$ decomposition, i.e. $J_{AB} = \{J_{ab}, J_{a*}\}$, where the extra dimension has been denoted by an asterisk. In terms of these new generators, the algebra reads

$$\begin{aligned}
[J_{ab}, J_{cd}] &= \delta_{bc} J_{ad} - \delta_{ac} J_{bd} + \delta_{ad} J_{bc} - \delta_{bd} J_{ac}, \\
[J_{ab}, J_{c*}] &= \delta_{bc} J_{a*} - \delta_{ac} J_{b*}, \\
[J_{a*}, J_{c*}] &= -\kappa J_{ac},
\end{aligned} \tag{48}$$

with[8]

$$\kappa = \begin{cases} 1 & SO(4) \\ 0 & ISO(3) \\ -1 & SO(3,1) \end{cases}. \tag{49}$$

The six Killing vectors of these algebras, expressed in spherical coordinates, are,

$$\begin{aligned}
J_1 &= J_{23} = \begin{pmatrix} 0 & 0 & -\cos(\varphi) & \cot(\theta)\sin(\varphi) \end{pmatrix}, \\
J_2 &= J_{31} = \begin{pmatrix} 0 & 0 & \sin(\varphi) & \cos(\varphi)\cot(\theta) \end{pmatrix}, \\
J_3 &= J_{12} = \begin{pmatrix} 0 & 0 & 0 & 1 \end{pmatrix}, \\
P_1 &= J_{1*} = \sqrt{1-\kappa r^2}\begin{pmatrix} 0 & \cos(\varphi)\sin(\theta) & \dfrac{\cos(\varphi)\cos(\theta)}{r} & -\dfrac{\sin(\varphi)}{t\sin(\theta)} \end{pmatrix}, \\
P_2 &= J_{2*} = \sqrt{1-\kappa r^2}\begin{pmatrix} 0 & \sin(\varphi)\sin(\theta) & \dfrac{\cos(\theta)\sin(\varphi)}{r} & \dfrac{\cos(\varphi)}{t\sin(\theta)} \end{pmatrix}, \\
P_3 &= J_{3*} = \sqrt{1-\kappa r^2}\begin{pmatrix} 0 & \cos(\theta) & -\dfrac{\sin(\theta)}{r} & 0 \end{pmatrix}.
\end{aligned} \tag{50}$$

Using Eq. (47), for the above Killing vectors, the most general connection compatible with the desired symmetries can be obtained [47], giving the components structure shown in Eq. (24).

Author details

Oscar Castillo-Felisola[1,2*], José Perdiguero[1] and Oscar Orellana[3]

1 Departamento de Física, Universidad Técnica Federico Santa María, Valparaíso, Chile

2 Centro Científico Tecnológico de Valparaíso, Valparaíso, Chile

3 Departamento de Matemáticas, Universidad Técnica Federico Santa María, Valparaíso, Chile

*Address all correspondence to: o.castillo.felisola@gmail.com

[8] The inhomogeneous algebra of $ISO(n)$ can be obtained from those of $SO(n+1)$ or $SO(n,1)$ through the Inönü-Wigner contraction [65].

References

[1] Einstein A. Zur allgemeinen relativitätstheorie. Sitzungsberichte der Preussischen Akademie der Wissenschaften. 1915;**1**:778

[2] Hilbert D. Nachrichten von der gesellschaft der wissenschaften zu göttingen, mathematisch-physikalische klasse. Die Grundlagen der Physik. (Erste Mitteilung). 1915;**1915**:395

[3] Will CM. The confrontation between general relativity and experiment. Living Reviews in Relativity. 2014;**17**:4. [1403.7377]

[4] Virgo and Ligo Scientific Collaboration, Abbott BP, et al. Observation of gravitational waves from a binary black hole merger. Physical Review Letters. 2016;**116**:061102. [1602.03837]

[5] Abbott BP et al. Gravitational waves and gamma-rays from a binary neutron star merger: Gw170817 and grb170817a. The Astrophysical Journal. 2017;**848**:L13

[6] Eddington AS. The Mathematical Theory of Relativity. London: Cambridge University Press; 1923

[7] Schrödinger E. Space-Time Structure. London: Cambridge University Press; 1950

[8] Bañados M, Ferreira PG. Eddington's theory of gravity and its progeny. Physical Review Letters. 2010;**105**: 011101

[9] Bañados M, Ferreira PG. Erratum: Eddington's theory of gravity and its progeny [Phys. Rev. Lett. 105, 011101 (2010)]. Physical Review Letters. 2014; **113**:119901

[10] Cartan E. Sur une généralisation de la notion de courbure de riemann et les espaces à torsion. Comptes Rendus Acad. Sci. Paris. 1922;**174**:593

[11] Cartan E. Sur les variétés à connexion affine et la théorie de la relativité généralisée (première partie). Annales Scientifiques de L'Ecole Normale Superieure. 1923;**40**:325

[12] Cartan E. Sur les variétés à connexion affine, et la théorie de la relativité généralisée (première partie) (suite). Annales Scientifiques de L'Ecole Normale Superieure. 1924;**41**:1

[13] Cartan E. Sur les variétés à connexion affine et la théorie de la relativité généralisée, (deuxième partie). Annales Scientifiques de L'Ecole Normale Superieure. 1925;**42**:17

[14] Kibble TWB. Lorentz invariance and the gravitational field. Journal of Mathematical Physics. 1961;**2**:212

[15] Hehl FW, McCrea JD, Mielke EW, Ne'eman Y. Metric affine gauge theory of gravity: Field equations, Noether identities, world Spinors, and breaking of dilation invariance. Physics Reports. 1995;**258**:1. [gr-qc/9402012]

[16] Mardones A, Zanelli J. Lovelock–Cartan theory of gravity. Classical and Quantum Gravity. 1991;**8**:1545

[17] Kijowski J. On a new variational principle in general relativity and the energy of the gravitational field. General Relativity and Gravitation. 1978;**9**:857

[18] Krasnov K. Non-metric gravity: A status report. Modern Physics Letters A. 2007;**22**:3013. [0711.0697]

[19] Krasnov K. Non-metric gravity. I. Field equations. Classical and Quantum Gravity. 2008;**25**:025001. [gr-qc/0703002]

[20] Krasnov K. Pure connection action principle for general relativity. Physical Review Letters. 2011;**106**:251103. [1103.4498]

[21] N. J. Popławski, A unified, purely affine theory of gravitation and electromagnetism. 0705.0351

[22] Popławski NJ. On the nonsymmetric purely affine gravity. Modern Physics Letters A. 2007;**22**:2701. [gr-qc/0610132]

[23] Popławski NJ. Affine theory of gravitation. General Relativity and Gravitation. 2014;**46**:1625. [1203.0294]

[24] Castillo-Felisola O, Skirzewski A. A polynomial model of purely affine gravity. Revista Mexicana de Fisica. 2015;**61**:421. [1410.6183]

[25] Castillo-Felisola O, Skirzewski A. Einstein's gravity from a polynomial affine model. Classical and Quantum Gravity. 2018;**35**:055012. [1505.04634]

[26] Azri H, Demir D. Affine inflation. Physical Review D. 2017;**95**:124007

[27] Azri H, Demir D. Induced affine inflation. Physical Review D. 2018;**97**:044025

[28] Azri H. Are there really conformal frames? Uniqueness of affine inflation. International Journal of Modern Physics D: Gravitation; Astrophysics and Cosmology. 2018;**27**:1830006

[29] Azri H. Cosmological implications of affine gravity [Ph.D. thesis]. İzmir Institute of Technology; 2018. 1805.03936

[30] Palatini A. Deduzione invariantiva delle equazioni gravitazionali dal principio di hamilton. Rendiconti del Circolo Matematico di Palermo. 1919;**43**:203

[31] Debever R, editor. Elie Cartan—Albert Einstein Letters on Absolute Parallelism 1929–1932. Princeton: Princeton University Press; 1979

[32] Curtright T. Generalized gauge fields. Physics Letters B. 1985;**165**:304

[33] Buchholz D, Fredenhagen K. Dilations and interaction. Journal of Mathematical Physics. 1977;**18**:1107

[34] McGady DA, Rodina L. Higher-spin massless S-matrices in four-dimensions. Physical Review D. 2014;**90**:084048. [1311.2938]

[35] Camanho XO, Edelstein JD, Maldacena J, Zhiboedov A. Causality constraints on corrections to the graviton three-point coupling. Journal of High Energy Physics. 2016;**2016**. [1407.5597]

[36] Nomizu K, Sasaki T. Affine Differential Geometry. Cambridge: Cambridge University Press; 1994

[37] Bryant RL. Symmetries of non-Riemannian curvature tensor (answer). MathOverflow

[38] Derdziński A. Classification of certain compact Riemannian manifolds with harmonic curvature and non-parallel Ricci tensor. Mathematische Zeitschrift. 1980;**172**:273

[39] Besse AL. Einstein Manifolds. Berlin: Springer; 2007

[40] Stephenson G. Quadratic lagrangians and general relativity. Nuovo Cimento. 1958;**9**:263

[41] Kilmister CW, Newman DJ. The use of algebraic structures in physics. In: Mathematical Proceedings of the Cambridge Philosophical Society. Vol. 57. Cambridge: Cambridge University Press; 1961. p. 851

[42] Yang CN. Integral formalism for gauge fields. Physical Review Letters. 1974;**33**:445

[43] Bourguignon J-P. Les variétés de dimension 4 à signature non nulle dont la courbure est harmonique sont d'einstein. Inventiones Mathematicae. 1981;**63**:263

[44] Derdziński A. Riemannian manifolds with harmonic curvature, in global differential geometry and global analysis 1984. In: Ferus D, Gardner RB, Helgason S, Simon U, editors. Lecture Notes in Mathematics. Vol. 1156. Berlin Heidelberg: Springer; 1985. p. 74

[45] Zanelli J. Private Communication

[46] Dray T. Differential Forms and the Geometry of General Relativity. London: CRC Press; 2014

[47] O. Castillo-Felisola, Gravity, ch. Beyond Einstein: A polynomial affine model of gravity, pp. 183–201. IntechOpen, Rijeka, 2018. 10.5772/intechopen.70951

[48] Weinberg S. The cosmological constant problem. Reviews of Modern Physics. 1989;**61**:1

[49] Markevitch M et al. Direct constraints on the dark matter self-interaction cross section from the merging galaxy cluster 1E 0657-56. The Astrophysical Journal. 2004;**606**:819

[50] van Dokkum P et al. A galaxy lacking dark matter. Nature. 2018;**555**:629

[51] Supernova Search Team Collaboration, Riess AG, et al. Observational evidence from supernovae for an accelerating universe and a cosmological constant. Astronomy Journal. 1998;**116**:1009. [astro-ph/9805201]

[52] Supernova Cosmology Project Collaboration, Perlmutter S, et al. Measurements of Ω and Λ from 42 high-redshift supernovae. The Astrophysical Journal. 1999;**517**:565

[53] Stein WA et al. Sage Mathematics Software (Version 8.2). The Sage Development Team; 2018

[54] Gourgoulhon E, Bejger M, et al. SageManifolds (Version 1.2). SageManifolds Development Team; 2018

[55] Gourgoulhon E, Bejger M, Mancini M. Tensor calculus with open-source software: The sagemanifolds project. Journal of Physics: Conference Series. 2015;**600**:012002. [1412.4765]

[56] Gourgoulhon E, Mancini M. Symbolic tensor calculus on manifolds: A sagemath implementation. Clinical Orthopaedics and Related Research. 2018. [1804.07346]

[57] Peeters K. Symbolic field theory with cadabra. Computeralgebra Rundbrief. 2007;**41**:16

[58] Peeters K. Introducing cadabra: A symbolic computer algebra system for field theory problems. hep-th/0701238

[59] Peeters K. Cadabra: A field-theory motivated symbolic computer algebra system. Computer Physics Communications. 2007;**176**:550

[60] Schwarzschild K. On the gravitational field of a mass point according to Einstein's theory. Sitzungsberichte der Königlich Preussischen Akademie der Wissenschaften zu Berlin, Physikalisch-Mathematische Klasse. 1916;**1916**:189. [physics/9905030]

[61] Yano K. The Theory of Lie Derivatives and its Applications. Vol. 3. Amsterdam: North-Holland; 1957

[62] Choquet-Bruhat Y, DeWitt-Morette C, Dillard-Bleick M. Analysis, Manifolds and Physics. Vol. 1 & 2. Amsterdam: North-Holland; 1989

[63] Nakahara M. Geometry, Topology and Physics. Bristol: Institute of Physics; 2005

[64] McInerney A. First Steps in Differential Geometry. Berlin: Springer; 2013

[65] Gilmore R. Lie Groups, Lie Algebras, and Some of their Applications. New York: Dover; 2005

Chapter 6

Cosmological Constant and Particle Masses in Conformal Quantum Gravity

Ho-Ming Mok

Abstract

It has been proposed that the equivalence principle of quantum gravity should be introduced as a fundamental symmetry in quantum gravity to reconcile quantum mechanics and general relativity. Such symmetry extends the equivalence principle of general relativity to the observer frames of reference which are in quantum mechanical motions. That means the quantum state of a particle is relative to the observer frame which can also be itself in a quantum mechanical state. As a consequence, all the physical laws apply not just to the frames of reference in any kind of motion as in the general relativity but also the same in the reference frames in quantum mechanical motions as well. The classical space-time concept therefore requires to be significantly modified. Because of such principle, the quantum gravity should be formulated in the quantum space-time-matter space with local conformal symmetry. In this book chapter, we explore the formulation of the quantum space-time-matter geometry with local conformal symmetry for discussing the relationship between the cosmological constant and quantum gravity as well as the mass spectrum of fundamental particles. The mathematical expressions of the fundamental particle masses and cosmological constant are discussed.

Keywords: cosmological constant, equivalence principle of quantum gravity, Higgs condensate, quantum space-time-matter geometry, local conformal symmetry

1. Introduction

The cosmological constant would be fundamentally related to the quantum nature of space-time. The author has proposed that the cosmological constant problem could be resolved by making the hypothesis that the space-time itself behaves as the phase of Higgs condensate (or say space-time condensate) and is discrete in nature. The estimated value of cosmological constant is in excellent agreement with the cosmological observations [1–4]. It has been further shown that the phase factor associated with the particle field in the discrete space-time could generate the CP violation in quark mixing system [5]. It is thus expected that the ultimate theory of quantum gravity would explain the cosmological constant problem.

However, there are fundamental inconsistencies between the general relativity and quantum mechanics in constructing a quantum theory of gravity. Actually, some important elements that are present in one theory are missing in the other. In the one hand, the general relativity does not involve the concepts of quantisation and probability as well as the uncertainty principle which are the important characteristics of quantum mechanics. On the other hand, quantum mechanics is not geometrical and there is no equivalence principle to make it independent with the quantum state of the observer frame of reference like the classical reference frame of general relativity. Such situation imposes fundamental difficulties in the unification of both theories. In order to achieve a more symmetrical treatment to bridge the gaps between both theories for their unification, it has been proposed that the equivalence principle of quantum gravity should be introduced in quantum gravity to reconcile the quantum mechanics and general relativity [6]. The equivalence principle of quantum gravity is that the laws of physics must be of such a nature that they apply to systems of reference in any kind of motion, both classical and quantum mechanical. Such symmetry extends the equivalence principle of general relativity to the observer frames of reference which are in quantum mechanical motions. That means the quantum state of a particle is relative to the observer frame of reference which can also be itself in a quantum mechanical state. As a consequence, all the physical laws apply not just to the frames of reference in any kind of classical motion as in general relativity but also the same in the reference frames in quantum mechanical motions as well. The classical space-time concept therefore requires to be modified significantly. Under such principle, the quantum gravity should be formulated in the quantum space-time-matter space with local conformal symmetry. The advantages of such treatment are that quantum mechanical motions of observers are introduced to extend general relativity from the classical to quantum mechanical domain. On the other hand, quantum mechanics can be made geometrical and relative to the quantum state of observer without any preferred frame. No preferred observer frame of reference is the essence of the principle of relativity.

In this book chapter, we explore the formulation of the quantum space-time-matter geometry with local conformal symmetry for discussing the relationship between the cosmological constant and quantum gravity as well as its connection with the Higgs condensate that would explain the nature of cosmological constant. Furthermore, such formulation implies that the mass spectrum of fundamental particles is related to the cosmological constant and the mathematical expressions of the fundamental particle masses and cosmological constant are discussed.

2. Quantum space-time-matter geometry

According to the principle of relativity, the space and time are defined by the coordinate system established respectively by the measuring-rods and synchronised clocks (or the equivalent measurement devices) at rest relative to the observer, which is known as the observer frame of reference, for the descriptions of physical events. In the theory of relativity, the measuring-rods and clocks are in classical motions and thus, within a specified measurement uncertainty, we can assume that their measuring results correspond to the points in the four-dimensional space-time coordinate system as space-time points. However, such classical concept of space-time can only be an approximation as the measuring-rods and clocks themselves are subject to quantum uncertainty and quantum mechanical motions, just the same as the observed matter particles. That means the measuring results of the measuring-rods and clocks should be probabilistic and subject to the uncertainty principle. When the interested scale of length and time are in microscopic level and the quantum uncertainties of the measuring-rods and clocks are large, we can imagine that no admissible coordinate system in classical sense can be established by such measuring-rods and clocks. On the other hand, if we require the quantum uncertainties of the measuring-rod and clock to be infinitely small, according to the uncertainty principle, the momentum and energy involved in the measurements, or say the observation or probing energy scale, should be infinitely large. However, when the energy and momentum is sufficiently large, according to the general relativity, a black hole with an event horizon comparable to the measuring scale will form and it will make the interested physical events hidden behind the event horizon and thus become inaccessible. That means it is not possible to achieve the measurement of space-time points. Such situations reveal that the classical space-time definition is obviously in trouble and modification to it is needed.

If we still adopt defining space-time by measuring-rods and clocks for describing the physical events, the quantum mechanical motions of such measuring-rods and clocks should be considered in the space-time definition. Therefore, the space-time cannot simply be a 4-dimensional coordinate system with space-time points but should be associated with the quantum states of measuring-rods and clocks specified by their state parameters. As the measuring-rod and clock are used for measuring the space and time, their quantum states should be characterised by the classical space-time vector $\vec{x}$; otherwise, they are not the suitable measuring devices for performing the space-time measurements. Furthermore, because of the constraint of measurement scale by the uncertainty principle as explained before, it is reasonable to introduce the quantum uncertainties associated with the space-time measurements as additional state parameters of the measuring-rod and clock. This treatment has the advantage that it is similar to specifying errors or uncertainties for experimental data but the major difference now is that the quantum uncertainty introduced should be an intrinsic spacetime property, rather than instrumental errors or uncertainties. This will lead to a very fundamental change on the spacetime concept. In this connection, let us denote the quantum state of the measuring-rod and clock, that is the quantum state of the spacetime by definition, as

$$\left|\vec{x}, \Delta\,\vec{x}\right\rangle \tag{1}$$

where $\vec{x}$ is the classical spacetime vector and $\Delta\vec{x}$ is the associated uncertainty. It should be noted that the above notation represents the quantum states of spacetime associated quantum uncertainties, which is the states of coordinate system established by measuring-rods and clocks, rather than the particle states. As the quantum uncertainties of the measuring-rod and clock depends upon their energy scales, the modified definition of spacetime should therefore be observation energy scale dependent.

In order to have a proper representation space for the quantum state of particle, the spacetime state should satisfy the following relation

$$\left|\vec{x} + \delta\,\vec{x}, \Delta\vec{x}\right\rangle = \left|\vec{x}, \Delta\vec{x}\right\rangle + \delta\left|\vec{x}, \Delta\vec{x}\right\rangle \tag{2}$$

The quantum state $|\alpha\rangle$ of a particle can be projected to such quantum spacetime state to become a wave function $\phi\left(\vec{x}, \Delta\vec{x}\right)$ as

$$\phi\left(\vec{x}, \Delta\vec{x}\right) = \left\langle \vec{x}, \Delta\vec{x}\,|\alpha\right\rangle \tag{3}$$

From the above expression, one may find that, although such expression of wave function in terms of state kets is somehow similar to the usual expression in the ordinary quantum mechanics, the major differences between them are that the spacetime uncertainty should be specified in the quantum spacetime representation for the particle quantum state and the spacetime states are in general not a spacetime point. Furthermore, in the ordinary quantum mechanics, the expression $\left\langle \vec{x} \,|\alpha\right\rangle$ represents the projection of the particle quantum state $|\alpha\rangle$ to its position eigenstate $\left|\vec{x}\right\rangle$. The position eigenstate is simply the particle position quantum state as measured by the classical frame of reference. Whereas, the expression in Eq. (3) extends such meaning to the measurement of the particle position quantum state by an observer in quantum mechanical motion. The projection in the ordinary quantum mechanics therefore becomes a special case of it. The above treatment introduces quantisation and probability to the spacetime concept as required in the actual situation of physical measurements in microscopic scale and also, the meaning of particle wave function is extended to become the observation of the particle quantum state in a specific quantum spacetime state.

We can imagine that the particle quantum state can be observed in another quantum spacetime state $\left|\vec{y}, \Delta\ \vec{y}\right\rangle$ which can be expressed in general as linear combination of the eigenstate of $\left|\vec{x}, \Delta\ \vec{x}\right\rangle$. We can therefore express the transformation as

$$\left\langle \vec{y}, \Delta\ \vec{y}\ |\alpha\right\rangle = \int \left\langle \vec{y}, \Delta\ \vec{y}\ |\ \vec{x}, \Delta\ \vec{x}\right\rangle \left\langle \vec{x}, \Delta\ \vec{x}\ |\alpha\right\rangle \tag{4}$$

The integration sign represents the summation or integration of all the possible eigenstates of $\left|\vec{x}, \Delta\ \vec{x}\right\rangle$. Actually, we can omit such notation, when summing or integrating the states, by identifying the operator $\left|\vec{x}, \Delta\ \vec{x}\right\rangle \left\langle \vec{x}, \Delta\ \vec{x}\right|$ as the internal index of the summation or integration. That is similar to the Einstein's convention of omitting the summation sign when summing the internal index of the product of contravariant and covariant tensor in the general relativity. The expression in Eq. (4) is directly analogous to representing the projection of four vectors in different coordinate systems, which is associated with different frame of references, in special relativity. The above treatment allows us to bring the general relativity closer to quantum mechanics in constructing the theory of quantum gravity.

In order to bring the quantum mechanics closer to the general relativity, we need to introduce the equivalence principle and geometrical concept to the ordinary quantum mechanics in the theory of quantum gravity. As mentioned above, in the relativity, the measuring-rod and clock at rest relative to the observer defines the observer frame of reference. However, the measuring-rod and clock are subject to quantum mechanical motion and described by the quantum spacetime state $\left|\vec{x}, \Delta\ \vec{x}\right\rangle$, which means the observer frame of reference in different quantum mechanical motions should be described by different quantum spacetime states. As the equivalence principle of general relativity requires that all the physical laws apply to the frames of reference in any kind of classical motion but actually the observer frames of references can be in quantum mechanical motions, it is therefore natural and reasonable to extend the equivalence principle to the observer frames of references in quantum mechanical motions. As a consequence, all the physical laws apply not just to the frames of reference in any kind of classical motion as in general relativity but also the same in the reference frames in quantum mechanical motions as well. We call this the equivalence principle of quantum gravity [6]. Since different quantum mechanical observer frames are described by different quantum spacetime states, the equivalence principle of quantum gravity implies that the general laws of nature are required to be expressed by equations which hold good for all systems of quantum spacetime states which are covariant under the transformation in Eq. (4). Actually, this is an extension of the general covariance from the classical coordinate system to the to the quantum spacetime space.

In the general relativity, under the equivalence principle, we can always find a free fall frame, which is local in spacetime, in which there is no gravity acting on the observed particles. Similarly, based on the equivalence principle of quantum gravity, we may anticipate that we can always find a "quantum free" frame for a matter particle in which there is no observed quantum effect on the observed particle. However, the major difference between such arguments is that all the matter particles with different masses have the same acceleration under the gravity, whereas the quantum mechanical motions depend on the energy and momentum of the particles. That means the "quantum free" frame for one particle may not be valid for another, especially when two particles of different energy and momentum are observed together under the same quantum spacetime state. In the general relativity, when a particle is moving in a gravitational field, its equation of motion is governed by the geodestic equation given by the extremum of the following action

$$S = \int ds \tag{5}$$

However, as we know in relativistic mechanics, the equation of motion of free particle is given by the least action principle on the following action

$$S = \int mds \tag{6}$$

The difference between them is that the rest mass is omitted in the particle action in general relativity. This is due to the fact that the inertial mass is the same as the gravitational mass under the equivalence principle of general relativity. However, when the quantum mechanical motion of matter particle is considered, the inertial mass cannot be neglected. We can understand this by considering the Schrödinger equation of a particle in gravitational field. As demonstrated by the COW experiment [7], the quantum interference pattern of neutrons induced by gravity depends upon the neutron inertial mass. Although it does not imply that the equivalence principle of general relativity is violated in the quantum mechanical particles under gravity, the experiment does show that the quantum mechanical motions of different particle masses under gravitational field cannot be made geometrical in spacetime coordinate system. This is another fundamental difficulty in unifying quantum mechanics with general relativity. In connection to this, the author has proposed that the spacetime should be merged with matter together to become the space-time-matter space in formulating the theory of quantum gravity [6]. But we cannot simply write the geometrical line element mds as the action for a particle in quantum gravity because of the quantum mechanical nature of matter particles as well as the above mentioned quantum spacetime concept. In this connection, let us introduce the quantum space-time-matter state $|\Psi\rangle$ by combining the quantum state of particle and spacetime as

$$|\Psi\rangle = |\phi\rangle \left| \vec{x}, \Delta\, \vec{x} \right\rangle \tag{7}$$

As the quantum space-time-matter space combines the quantum states of the matter particle and the spacetime, it should therefore be complex in nature. The quantum space-time-matter space also allows different quantum spacetime for describing different quantum particle. Then, by introducing the equivalence principle of quantum gravity, we can make the quantum mechanics become a geometrical theory in quantum space-time-matter space. Under the equivalence principle of quantum gravity, all the physical laws apply to the observer frames of reference in any classical and quantum mechanical motions, which means there is no preferred observer frame of reference in any classical and quantum mechanical motions. The particle states and the equation of motion should therefore be equally good for any observer frame of references with the same physical laws. As the general change of observer frame of reference should be associated with the change of the spacetime state, this requires the general covariance of the physical laws under the transformation of the spacetime state in Eq. (4). In the general relativity, the physical observations in accelerating observer frames are associated with the generalised spacetime coordinate system. The general covariance on spacetime coordinate system allows us to choose a local observer frame which is an inertial frame such that there is no acceleration on the observed particles. This is what we call the "free fall frame". The gravitational field, which provides universal acceleration on all matter particles, corresponds to the curvature of spacetime. Analogously, the extension of the physical observations in quantum observer frames is associated with the generalised quantum spacetime states system. The extended general covariance on the quantum spacetime states system allows us to choose a local quantum observer frame such that there is no observed quantum motion on the observed particles. We may call this the "quantum free" frame. The quantum mechanical motions then correspond to the curvature of the quantum space-time-matter space.

The equivalence principle of quantum gravity and the quantum spacetime concept therefore allows the interesting physical results that the quantum mechanical motion of a particle depends upon the quantum mechanical motion of the observer frame. That means the quantum mechanical motion is relative in nature. This can be further explained by the simple argument as follows. Let us suppose that there is an observer A and, with reference to its frame of reference, the observer B and observe C are in the same quantum mechanical states. Then, the state of observer C as observed by the observer B should be the same as the state of observer B as observed by the observer C since their states are the same and therefore should be symmetrical to each other. Analogous to the relativity that the observer is assumed to be at rest relative to the coordinate system, if we assume that the observer C can always find an appropriate energy scale to establish a coordinate system with a specified uncertainty that there is no observable quantum effect between the spacetime coordinates in the coordinate system. We can say that the space-time coordinates are free of quantum effect (mathematically speaking, this means that the observer as observed itself is an identity). Therefore, due to the symmetry between observer B and C, the observer B should be in a "quantum free" state as observed by observer C in such "quantum free" coordinate system and vice versa. As a result, we can always establish a "quantum free" frame for a quantum particle by choosing the quantum frame of reference at the same state of the observed particle.

As the quantum mechanical motion of a particle and the quantum spacetime state depends upon their energy scales, or say the term $mc/\hbar$, which means the changing of quantum spacetime state could be associated with the change of energy scale. As we understand in the relativistic quantum mechanics, the normalisation factor, say N, of a quantum state is related to the energy scale so that any change of the energy scale should be associated with the change of N. That means the transformation in Eq. (4) between different quantum spacetime states is not necessary a unitary transformation, which preserves the total probability of quantum states, due to the change of normalisation in different quantum observer frames. Such change of the normalisation of the quantum states, no matter the particle state or the quantum spacetime state, mathematically acts like applying a conformal factor to the state as the conformal transformation. Furthermore, as the quantum spacetime states are in the complex domain and the particle density is related to the modulus of N, the most general transformation between different quantum spacetime states could also involve a complex phase factor $e^{i\delta}$ multiplying to the conformal factor. This phase factor could have important physical meaning to the CP-violation problem [5] but we will limit our discussions in this chapter without considering such phase factor. As the quantum spacetime state is allowed to be changed with the spacetime variables, the conformal transformation should be local in nature. Since the equivalence principle of quantum gravity requires that there is no preferred quantum spacetime state for observation, that means the equation of motion should then be invariant under the local conformal transformation and the quantum space-time-matter space should therefore possess the local conformal symmetry.

In some sense, such local conformal symmetry behaves as a kind of gauge symmetry on the quantum space-time-matter space, rather than only on the spacetime. The local conformal transformation in quantum space-time-matter space acts on the quantum space-time-matter metric whereas the conformal transformation in spacetime acts on the spacetime metric. This make the local conformal transformation in the quantum space-time-matter space something different from the usual conformal transformation in spacetime. In contrast with the general relativity, of which its equivalence principle implies the general covariance of system of spacetime co-ordinates for expressing the physical laws, the equivalence principle of quantum gravity implies the general covariance with scale change of the system of co-ordinates of space-time-matter space. The incorporation of the local conformal symmetry in the theory make it behaves as Weyl like geometry but now with the spacetime replaced by the quantum space-time-matter space.

Analogous to the general relativity, in constructing the geometry of general covariance for a curved quantum space-time-matter space, we may consider a general small line element on a small region of quantum space-time-matter space. Let us define the length of such line element dL as the inner product of $\delta|\Psi\rangle$ in a small region of quantum space-time-matter space as

$$dL^2 = \delta|\Psi\rangle \cdot \delta|\Psi\rangle = \delta\left(|\phi\rangle\left|\vec{x}, \Delta\vec{x}\right\rangle\right) \cdot \delta\left(|\phi\rangle\left|\vec{x}, \Delta\vec{x}\right\rangle\right) \quad (8)$$

where the inner product can be defined as the usual inner product used for quantum mechanical states and thus dL^2 has a real value. In fact, it is reasonable to take real value for the inner product of the projection of a state onto itself. However, because of the local conformal symmetry, dL^2 should not be an invariant under the transformation in Eq. (4) due to the change of the observer frame of reference. Let us also introduce the operator $\widehat{\delta\vec{x}_{\Delta x}}$, which can extract the spacetime length from the quantum spacetime state, for small changes on spacetime state with the property as

$$\widehat{\delta\vec{x}_{\Delta x}}\delta\left|\vec{x}, \Delta\vec{x}\right\rangle = \delta\vec{x}_{\Delta x}\delta\left|\vec{x}, \Delta\vec{x}\right\rangle \quad (9)$$

Suppose that we do not consider any mixing between the quantum matter particle state and the quantum spacetime state, of which they respectively correspond to the object and observer, by any symmetry operation, we may define another line element dl of quantum space-time-matter space, which combines the inner product of the quantum space-time-matter state with the extracted spacetime vector, as

$$dl^2 = \delta|\phi\rangle\widehat{\delta\vec{x}_{\Delta x}\delta}\left|\vec{x}, \Delta\,\vec{x}\right\rangle \cdot \delta|\phi\rangle\widehat{\delta\vec{x}_{\Delta x}\delta}\left|\vec{x}, \Delta\,\vec{x}\right\rangle \tag{10}$$

The element dl^2 is also real and can be expressed in generalised coordinates ξ_m^μ in the quantum space-time-matter space as

$$dl^2 = G_{\mu\nu}^{mn} d\xi_m^\mu d\xi_n^\nu \tag{11}$$

where $G_{\mu\nu}^{mn}$ is the combined metric of the quantum space-time-matter space and the extracted spacetime length and the index μ and ν are the usual spacetime index in general relativity while m and n are the index associated with the inner product of the quantum spacetime and matter states. We have assumed the expansion of the quantum space-time-matter states by the discrete eigenstates such that the discrete indices can be used above as analogous to the general relativity. But actually, expansion on continuous eigenstates can be used with the integration, rather than summation, as well in the expression without changing the formulation. If we formulate the curved quantum space-time-matter space with local conformal symmetry, let us define the local conformal operator Ω on the quantum space-time-matter state as

$$\delta|\Psi\rangle \to \Omega\delta|\Psi\rangle \tag{12}$$

As mentioned, we will not discuss the phase factor which may appear with the conformal operator in the transformation of the quantum spacetime. In view of the local conformal symmetry, we can make use of the Weyl geometry for the quantum space-time-matter space. The closest admissible action analogously to the action for Einstein equation under the local conformal symmetry would be taken as [8].

$$S = \int \mathfrak{R}^2 \sqrt{-G} d^n \xi \tag{13}$$

where $\mathfrak{R}$ is the curvature scalar defined for the quantum space-time-matter space in analogous to Weyl geometry and n is the number of dimensions of the quantum space-time-matter space. This is a generalised action of quantum gravity in the formulation of the quantum space-time-matter space of which the generalised coordinates of spacetime and matter are combined together. Such action determines the combined scale of the matter state and the associated spacetime energy scale. The variation of the action gives

$$\delta S = \delta \int \mathfrak{R}^2 \sqrt{-G} d^n \xi = \int \left(2\mathfrak{R}\delta\mathfrak{R}\sqrt{-G} + \mathfrak{R}^2 \delta\sqrt{-G}\right) d^n \xi = 0 \tag{14}$$

As in Weyl geometry, the curvature scalar can be expressed as

$$\mathfrak{R} = \tilde{\mathfrak{R}} - (n-1)(n-2)\frac{(\partial\Phi)^2}{(\Phi)^2} + 2(n-1)\frac{1}{\sqrt{-G}}\left(\sqrt{-G}\frac{\partial^\alpha \Phi}{\Phi}\right)_{|\alpha} \tag{15}$$

where $\tilde{\mathfrak{R}}$ is the metric $G_{\mu\nu}^{mn}$ dependent part of the curvature scalar in Weyl geometry. The last term is the boundary term which can be neglected in the variation of the action. Actually, the Φ field is the scale of metric $G_{\mu\nu}^{mn}$ which determines the relative quantum effect of a particle as observed in a frame of reference in quantum mechanical motion. If we impose the gauge $\mathfrak{R} \sim 4\Lambda$ for fixing the scale of the quantum space-time-matter space, where Λ is a constant, Eq. (13) becomes

$$S = \int \left(\tilde{\Re} - 2\Lambda - (n-1)(n-2)\frac{(\partial\Phi)^2}{(\Phi)^2} \right) \sqrt{-G} d^n \xi \tag{16}$$

In order to make the expression of the action simpler, as analogous to the Weyl geometry of spacetime, we can perform a local conformal transformation on the metric as $G_{\mu\nu}^{mn} \to \Phi^2 G_{\mu\nu}^{mn}$ to remove the Φ field in the action before imposing the gauge $\Re \sim 4\Lambda$. The action can be reduced to

$$S = \int \left(\tilde{\Re} - 2\Lambda \right) \sqrt{-G} d^n \xi \tag{17}$$

The equation shows that the term Λ, which determines the scale of the quantum space-time-matter space, behaves mathematically like the cosmological constant in general relativity but it is now in the quantum space-time-matter space rather and its physical meaning is different.

We can then connect the formulation of quantum space-time-matter space to the general relativity by further specifying the relationship between the spacetime element ds in the general relativity and the quantum spacetime. Also, for simplicity, we will assume that the matter field is a scalar field. In fact, the argument can be extended to the fermionic field as well as the tetrad e_μ^I formalism of gravity [9, 10]. Firstly, let us write

$$\langle \delta\phi^2 ds^2 \rangle = \delta|\phi\rangle \widehat{\delta\vec{x}_{\Delta x}} \delta\left|\vec{x}, \Delta\ \vec{x}\right\rangle \cdot \delta|\phi\rangle \widehat{\delta\vec{x}_{\Delta x}} \delta\left|\vec{x}, \Delta\ \vec{x}\right\rangle \tag{18}$$

where $\langle \delta\phi^2 ds^2 \rangle$ is the combined inner product of the quantum space-time-matter state with the extracted spacetime vector for a small region of quantum space-time-matter space. It acts like the expectation value of quantum probability weighted spacetime and matter element and then Eq. (10) above becomes

$$dl^2 = \langle \delta\phi^2 ds^2 \rangle \tag{19}$$

If we consider the special case that the quantum-space-time-matter space is a linear space, which is the metric $G_{\mu\nu}^{mn}$ does not vary with the associated generalised coordinates ξ_m of the quantum space-time-matter space, the above expression can be simplified as

$$dl^2 = \langle \phi^2 ds^2 \rangle \tag{20}$$

Actually, we can express the curved quantum spacetime in generalised coordinates analogous to Weyl geometry as

$$\langle \phi^2 ds^2 \rangle = \left\langle \phi^2 \overline{g}(\Delta x)_{\mu\nu} \right\rangle dx^\mu dx^\nu \tag{21}$$

where $\left\langle \phi^2 \overline{g}(\Delta x)_{\mu\nu} \right\rangle$ is the quantum probability weighted spacetime metric, which is specified with measurement uncertainty, combined with the matter field. If the quantum spacetime state is the eigenstate of the $\overline{g}(\Delta x)_{\mu\nu}$, that is the observer frame and observation energy scale for the quantum spacetime state is the same as the spacetime metric specified with measurement uncertainty, Eq. (20) can be written as

$$dl^2 = \left\langle \phi^2 \overline{g}(\Delta x)_{\mu\nu} \right\rangle dx^\mu dx^\nu = \langle \phi^2 \rangle \overline{g}(\Delta x)_{\mu\nu} dx^\mu dx^\nu \tag{22}$$

As we assume that the quantum space-time-matter space is a linear space, the metric $G^{mn}_{\mu\nu}$ does not vary with ξ_m and it is not necessary to consider the variation of the ξ_m coordinates in the action in Eq. (17). The Weyl geometry of the quantum space-time-matter space can be reduced to that only the curvature on the spacetime metric is described. The number of dimension n of the action in Eq. (17) can therefore be reduced to four. Although the equation resembles the Einstein action with the cosmological constant in general relativity, it actually describes the dynamics of the quantum space-time-matter space rather than simply the spacetime. The dependence on the quantum probability and uncertainty parameter of $\left\langle \phi^2 \bar{g}(\Delta x)_{\mu\nu} \right\rangle$ provides a freedom for the spacetime metric to change scale with the matter field. For instance, when we need to probe into the microscopic scale by using high energy scale observation, the metric will change due to the uncertainty required changes. This let us to connect the metric in macroscopic scale with the microscopic scale observation. Physically, it means that when the spacetime observation is performed by quantum mechanical devices, the probability of observation of existence of spacetime state should be taken into account in the metric of quantum spacetime. This echoes with the mentioned requirements of introduction of the quantum uncertainty into the spacetime states when we need to consider the quantum mechanical nature of measuring-rods and clock to bring classical general relativity closer to quantum mechanics. One may find that when the probability of quantum spacetime metric tends to unity, $\bar{g}(\Delta x)_{\mu\nu} = g_{\mu\nu}$, which is the metric of the general relativity. For simplicity, let us write $\bar{g}(\Delta x)_{\mu\nu}$ as $\bar{g}_{\mu\nu}$. Then, the action in Eq. (17) now can be expressed in terms of $\bar{g}_{\mu\nu}$ and ϕ as

$$S = \int \left(\phi^2 \bar{R} - 2\Lambda\phi^4 + 6(\partial\phi)^2 \right) \sqrt{-\bar{g}} d^4x \tag{23}$$

The action behaves like the conformal coupling of a scalar field to the gravitation field. Actually, we can divide the action by the factor 1/12 to align the term $(\partial\phi)^2$ as the kinetic term of the ϕ field and gives

$$S' = \int \left(\frac{1}{12}\phi^2 \bar{R} - \frac{1}{6}\Lambda\phi^4 + \frac{1}{2}(\partial\phi)^2 \right) \sqrt{-\bar{g}} d^4x \tag{24}$$

If the Ricci scalar $\bar{R}$ of the spacetime is a constant, we will show that it relates to the mass term of the scalar matter field. Applying the spontaneous symmetry breaking on the ϕ field of such Lagrangian, the minimum of the potential gives the relation

$$\frac{\bar{R}}{4\Lambda} = \phi_0^2 \tag{25}$$

We later will know that selecting the value of ϕ_0 determines the relative scale between the matter field and spacetime metric. In fact, when selecting the scale in the variation of the Weyl action in Eq. (13), the local conformal symmetry of the quantum space-time-matter space is broken and the combined scale between the matter density and the four dimensional spacetime energy scale is fixed. But it does not mean that the local conformal symmetry on the spacetime and matter field which relates to the changing energy scale of observation is broken and actually, it is not. Let us define the conformal operator ω, which is associated with the change of scale of observation, in the spacetime on the metric and scalar field as

$$\bar{g}_{\mu\nu} \to \omega^2 \bar{g}_{\mu\nu} \quad (26)$$

and

$$\phi \to \omega^{-1} \phi \quad (27)$$

As mentioned, we will not discuss the phase factor which may appear with the conformal operator in the transformation of the quantum spacetime. We may find that the conformal transformation, which is the change of scale of observation, in spacetime does not change the action in Eq. (24), which is a conformal invariant. Actually, the change of scale of observation should preserve the uncertainty relation $\Delta x \Delta p \sim \hbar$, since such change of scale is governed by the uncertainty relation. The local conformal symmetry of spacetime is broken when the scale of observation is fixed. This is the case when we apply the spontaneous symmetry breaking condition in Eq. (25) to the action. One may be aware that we need to fix two degree of freedoms in determining the scale of the whole theory in quantum gravity. This point is very important since we will find that we can change the scale in one space and then compensate by the other to make it invariant. This property let the quantum gravity possess an interesting double conformal structure. Actually, the selection of scale in the variation of the Weyl action determines the scale of Λ. Then with the relationship of the spontaneous symmetry breaking in Eq. (25), all three parameters $\bar{R}$, Λ and ϕ_0 can be fixed so as the whole scale of the observer and matter space. Actually, if the electroweak energy density scale is chosen for $\Lambda\phi^4$ or $\bar{R}\phi^2$ in the variation of the Weyl action, the equation becomes a Higgs potential like Lagrangian. That means it is possible to interpret the Higgs potential as the broken Weyl action of the quantum space-time-matter space with an observation scale dependent metric. It provides the physical explanation of the Higgs potential and its relationship with quantum gravity. Furthermore, the interesting thing is that even the energy density scale is fixed at electroweak scale, we still have the freedom to choose the scale of observation by the said conformal transformation on spacetime. This will lead to the relationship between the cosmological constant and Higgs potential as well as the fundamental particle masses as discussed in the next section.

3. Fundamental particle masses

If we consider that the conformal symmetry is spontaneously broken in the action of Eq. (24), just as breaking the gauge symmetry in Higgs mechanism, we may consider the shift field h around the minimum of the potential as $\phi = \phi_0$. If we first consider the symmetry breaking at an energy scale $\phi_0^2 = 6m'^2$ related to the fundamental interaction, say electroweak scale, the equation becomes

$$S = \int \left(\frac{1}{2} m'^2 \bar{R} - 3\Lambda m'^4 - \frac{1}{2}(\partial h)^2 + \frac{1}{2}\bar{R}h^2 + \ldots \right) \sqrt{-\bar{g}} d^4x \quad (28)$$

If we now change the scale of observation by applying conformal transformation on the spacetime as

$$\bar{g}_{\mu\nu} \to \left(\frac{m'}{M_p} \right)^2 \bar{g}_{\mu\nu} \quad (29)$$

and

$$\phi \rightarrow \left(\frac{m'}{M_p}\right)^{-1}\phi \tag{30}$$

where the mass scale M_p is reduced Planck mass. Then we can write the equation as

$$S = \int \left(\frac{1}{2}M_p^2\overline{R'} - 3\Lambda M_p^4 - \frac{1}{2}(\partial h')^2 + \frac{1}{2}\overline{R'}h'^2 + \ldots\right)\sqrt{-\overline{g}}d^4x \tag{31}$$

We have to note that the Planck mass is an energy scale introduced here to the Lagrangian through the relative ratio with m' in the conformal transformation. As a result, the action become the gravitational action with cosmological constant and the excitation field become the matter field that coupled to the gravitational field and the square of Planck mass become the coupling constant between matter and gravity. The term $\overline{R'}h'^2$ acts as the mass term of the excitation field h. Unexpectedly, it also indicates that the spontaneous symmetry breaking on electroweak energy scale is associated with the symmetry breaking that generate the gravitational action. Since $\overline{R'}$ is the conformal transformation of the related to the vacuum energy of spacetime in empty space and since $\overline{R}\sqrt{-\overline{g}}$ should be transformed as $\overline{g}$ and therefore $\overline{R'}$ should be equal to $(m'/M_p)^2\overline{R}$ which is the particle mass square term. This implies the existence of a particle with mass $(m'/M_p)\overline{R}^{\frac{1}{2}}$. The calculation results mean that the particle mass of the matter field is related to the vacuum energy of the spacetime, which is the cosmological constant.

On the other hand, due to the double conformal symmetry structure of the quantum space-time-matter space, we may change the observer frame of reference for observing such particle and this will lead to the change on the broken scale of the quantum space-time-matter space. Recalling that the scale of the quantum space-time-matter space is fixed by the gauge $\Re \sim 4\Lambda$ and let the factor for the associated scale change is Ω^4, the scale of Λ is therefore changed as

$$\Lambda \rightarrow \Omega^4\Lambda \tag{32}$$

The scale change of Λ implies that the spontaneous symmetry breaking condition in Eq. (25) requires to be varied so that the mass value $(m'/M_p)\overline{R}^{\frac{1}{2}}$ in Eq. (31) of the said particle will be changed also. However, we still have not established the reference for the values of mass and energy and, actually, mass value is just the mass ratio relative to the mass standard reference which could fundamentally be the Planck mass M_p. Recalling that, in Eq. (31), the Planck mass is introduced only through the conformal transformation of spacetime. That means, in changing of frame of reference, the Planck mass is required to be re-determined in the new observer frame of reference. The freedom of choosing the observation scale by conformal transformation on the spacetime and matter field in the new frame of reference let us have the freedom to do the transformation such that the mass value as reference to the mass standard reference of a particle remained unchanged. This point is important in the equivalence principle of quantum gravity since it ensures that there is no preferred observer frame of reference such that the observer is not able to distinguish the nature of the frames of reference that he is sitting in for making his observations and measurements. As the mass value is determined by the ratio $\overline{R}\sqrt{-\overline{g}}/\phi^2$, for the mass remains unchanged in the new frame, we need to have

$$\frac{\overline{R}\sqrt{-\overline{g}}}{\phi^2} = \frac{m'}{M_p} = unchanged \tag{33}$$

As under the change of scale in Eq. (31), the term $\overline{R}\sqrt{-\overline{g}}$ will be transformed as

$$\frac{\overline{R}\sqrt{-\overline{g}}}{\phi^2} \rightarrow \frac{\Omega^4\overline{R}\sqrt{-\overline{g}}}{\phi^2} \tag{34}$$

and the conformal transformation ω on the spacetime and matter field induce the transformation as

$$\frac{\Omega^4\overline{R}\sqrt{-\overline{g}}}{\phi^2} \rightarrow \frac{\omega^4\Omega^4\overline{R}\sqrt{-\overline{g}}}{\phi^2} \tag{35}$$

Therefore, the unchanged mass ratio requires that $\omega = 1/\Omega$. Once the particle mass value remains unchanged in the transformed frame and since the observer is not able to distinguish the frames of reference that he makes the observations and measurements, in order to explore the structure of the quantum space-time-matter space under broken symmetry, we can repeatedly perform the above processes of changing observer frame such that in every changed frame the mass of the particle are still equal to m'/M_p, although the M_p varies with the transformed frame of reference. We can therefore find that a set of scale factors is allowed for the quantum space-time-matter space scale

$$(\Omega)^{2n} = \left(\frac{m'}{M_p}\right)^2 \Rightarrow \Omega = \left(\frac{m'}{M_p}\right)^{\frac{1}{n}} \tag{36}$$

where n is the number of the said transformation to achieve the mass factor m'/M_p in a frame. The relation means that, for observing a particle with mass factor m'/M_p in a frame, it is possible that it can be due to a number of conformal transformations on the quantum space-time-matter space to bring to the observed mass value in a frame of reference. That means, in an observer frame of reference, there is a set of possible scale factors Ω for the broken scale of quantum space-time-matter space. As in an observer frame of reference the Planck mass is fixed, the set of possible broken scales of quantum space-time-matter space implies that there is a set of possible mass states with the mass factors as in Eq. (36). Suppose $m'/M_p \sim 10^{-15}$, that is $m' = 2.435\,TeV$, a set of mass factors can be found for different values of n, for instance n = 1 to 5, as follows

$$\Omega_{m1} = \frac{m'}{M_p} = \frac{2.435\,GeV}{2.435 \times 10^{18}\,GeV} = 10^{-15} \tag{37}$$

$$\Omega_{m2} = \left(\frac{m'}{M_p}\right)^{\frac{1}{2}} = \left(\frac{2.435\,GeV}{2.435 \times 10^{18}\,GeV}\right)^{\frac{1}{2}} = 3.16 \times 10^{-8} \tag{38}$$

$$\Omega_{m3} = \left(\frac{m'}{M_p}\right)^{\frac{1}{3}} = \left(\frac{2,435\,GeV}{2.435 \times 10^{18}\,GeV}\right)^{\frac{1}{3}} = 1.0 \times 10^{-5} \tag{39}$$

$$\Omega_{m4} = \left(\frac{m'}{M_p}\right)^{\frac{1}{4}} = \left(\frac{2,435\,GeV}{2.435 \times 10^{18}\,GeV}\right)^{\frac{1}{3}} = 1.778 \times 10^{-4} \tag{40}$$

$$\Omega_{m5} = \left(\frac{m'}{M_p}\right)^{\frac{1}{5}} = \left(\frac{2,435\,GeV}{2.435 \times 10^{18}\,GeV}\right)^{\frac{1}{5}} = 1 \times 10^{-3} \tag{41}$$

As mentioned above, since the particle mass m_i is equal to $\Omega_{mi}\overline{R}^{-\frac{1}{2}}$, if $\overline{R}^{-\frac{1}{2}} = 10\,TeV$, the following mass values m_i can be found and when compared with the lepton masses data of the Particle Data Group [11]:

$$m_1 = 10^{-15}\overline{R'}^{-\frac{1}{2}} = 10^{-15} \times 10\,TeV = 0.01\,eV \sim m_\nu? \tag{42}$$

$$m_2 = 3.16 \times 10^{-8}\overline{R'}^{-\frac{1}{2}} = 3.16 \times 10^{-8} \times 10\,TeV = 316.23\,keV \sim m_e \tag{43}$$

$$m_3 = 1.0 \times 10^{-5}\overline{R'}^{-\frac{1}{2}} = 1.0 \times 10^{-5} \times 10\,TeV = 100\,MeV \sim m_\mu \tag{44}$$

$$m_4 = 1.778 \times 10^{-4}\overline{R'}^{-\frac{1}{2}} = 1.778 \times 10^{-4} \times 10\,TeV = 1.778\,GeV \sim m_\tau \tag{45}$$

$$m_5 = 1 \times 10^{-3}\overline{R'}^{-\frac{1}{2}} = 1 \times 10^{-3} \times 10\,TeV = 10\,GeV \tag{46}$$

where m_ν, m_e, m_μ and m_τ are respectively the mass of neutrinos, electron, muon and tau lepton. Although the mass values of the three neutrino mass states are so far not experimentally determined, the mass value of m_1 state is consistent with the cosmological constraint on the sum of the neutrino masses. The actual neutrino masses might not be a necessary degenerate as the radiative correction of different neutrino mass states is not considered in the above calculation and that would lead to small differences between the actual neutrino masses and the calculated mass state value. For m_2 state, it is of the same order of magnitude of the electron mass. The difference between them is about 38% and could be due to the radiative correction to the QED vacuum. When the value of the mass states become relatively large, we find that the mass of m_3 state is equal to the experimental mass value of muon up to about 5% and the m_4 state is even equal to the experimental mass value of tau lepton up to about 0.06%. For the mass states with $n>4$, there are many so far not experimentally observed particle states. They could be the dark matter particles or the mass resonance states. For instances, the mass value of m_5 state is consistent with the dark matter candidate with a mass of 10 GeV as proposed by some researchers [12].

In fact, the above argument can be extended to the case that n is allowed to be half-integer, in the formulation of the gravitational fields by tetrads e^I_μ for considering its coupling with fermions. In such case, we can find more mass states which are associated with half integer n values, for instance, $n = 1.5$, $n = 2.5$ etc. For $n = 4.5$ and $n = 8.5$, we can find that the mass value are respectively $m_{4.5} = 4.64GeV$ and $m_{8.5} = 171.9GeV$. The values are very close to the mass values of bottom and top quarks. That means the half integer values of n may somehow correspond to the quark masses. However, it is possible that, due to the QCD vacuum, the calculated values of the quark masses which are comparable with the QCD vacuum energy would have greater discrepancy with the actual current quark masses.

4. Cosmological constant

As we explained before, Eqs. (24) and (31) is related by the conformal transformation of spacetime and the matter field and spontaneous symmetry breaking which is determined by the observation scale. Actually, Eq. (24) resembles the following action of the Higgs potential when $\overline{R'}$ is a constant in one hand

$$S' = \int \left(\frac{1}{2}\mu^2\phi^2 - \frac{1}{4}\lambda\phi^4 + \frac{1}{2}(\partial\phi)^2\right)\sqrt{-\overline{g}}d^4x \tag{47}$$

On the other hand, as shown above, it can become the action of gravitational field coupled to matter field after the local conformal transformation and spontaneous symmetry breaking as in Eq. (31). By comparing Eqs. (24) and (47), we can relate the coefficients between them as follows

$$\frac{\overline{R}}{12} = \frac{\mu^2}{2} \tag{48}$$

$$\frac{\Lambda}{6} = \frac{\lambda}{4} \tag{49}$$

Furthermore, by comparing Eq. (31) with the gravitational action, we can find that

$$3\Lambda M_p^2 = \Lambda_{cc} \tag{50}$$

where Λ_{cc} is the cosmological constant. By eliminating Λ in the above equations, we get

$$\Lambda_{cc} = 4.5\lambda M_p^2 \tag{51}$$

The calculated cosmological constant is of the order of Planck scale. It is due to the fact that the observation scale is in the microscopic scale or say high energy scale of which the metric $\overline{g'}_{\mu\nu}$ is quantum mechanically uncertain in nature. Thus, it is not the scale of our cosmological observation. Actually, we can apply a conformal transformation on the quantum space-time-matter space metric to change the observer frame of reference which is under the same factor as the one in obtaining the fundamental particle masses but with $n = 2$ as

$$\Omega = \left(\frac{m'}{M_p}\right)^2 \tag{52}$$

Since the cosmological constant transforms as the square of the metric, that is conformal factor Ω^4, with respect to the conformal factor of Ω^2 on the metric, the cosmological constant value becomes

$$\Lambda_{cc} = 4.5\lambda \frac{m'^4}{M_p^2}\left(\frac{m'}{M_p}\right)^4 \tag{53}$$

By putting $\lambda = 0.258$, which is based on the 125 GeV Higgs particle mass and 246 GeV electroweak VEV value,

$$\Lambda_{cc} = 6.49 \times 10^{-66}\, eV^2 \tag{54}$$

This calculated cosmological constant value is in very good agreement with the observation value of $4.33 \times 10^{-66}\, eV^2$ of Planck CMB probe with just about 50% difference when connecting to the fundamental particle masses.

5. Discussion and conclusions

We have formulated the quantum space-time-matter geometry with the equivalence principle of quantum gravity for discussing the relationship between the cosmological constant and quantum gravity as well as the mass spectrum of fundamental particles. Because of the freedoms allowed for changing the quantum observer frames under the equivalence principle of quantum gravity and the energy scale specific observation required under the quantum properties of the spacetime measuring devices, the quantum space-time-matter space possesses a double conformal symmetry geometrical structure. In such structure, we need two parameters to determine the scale of quantum space-time-matter space so as to describe the physical world.

In our calculation, it is fascinating that, once such parameters are determined, the associated conformal factor $\Omega = \left(\frac{m'}{M_p}\right)$ with different exponent n values gives the fundamental particle mass values and the observed cosmological constant value with very good agreement. It indicates that the fundamental particle masses is connected to the cosmological constant and some fundamental physical meaning are behind such factor, particularly the meaning of the value of 2.435 TeV. It also demonstrates that the quantum space-time-matter space has a complicated conformal structure which is related to the fundamental particles when applying the equivalence principle of quantum gravity to it.

Actually, we can extend the theory by incorporating the gauge symmetry of the fundamental interactions to it so as to discuss the relationship between the mass spectrum of fundamental particles and the standard model under the conformal nature of quantum gravity. As indicated in our calculation, it can be anticipated that the gauge symmetry operation for the flavour states of lepton would in general not be commutable with the conformal symmetry operation of the mass states. The mass states are under a conformal symmetry of the quantum space-time-matter space whereas the flavour states are under the gauge symmetry. This is consistent with the fact that the neutrino flavour eigenstates are not the same as their mass states. Our formulation indicates that the mass states can even be changed from one to another when observing under different observer frame of reference in quantum mechanical motions. As we can see, the underlying symmetry that associated with particle masses would require the combination of the local conformal symmetry with the gauge symmetry into the local conformal gauge symmetry.

Furthermore, the above mass formula allows the existence of some mass states that are so far not experimentally observed. This might provide new opportunities for discussing whether such mass states are related to the dark matter as well as the possibilities of discovering such particles experimentally, although some of the mass states may be forbidden by some so far unknown physical rules, just as the forbidden rules of atomic spectra.

One of the key underlying physical meanings of the above calculation is that there is no absolute existence of spacetime. Similar to the quantum nature of the matter fields, the existence of spacetime is probabilistic in nature. And it is due to such reason, the vacuum energy can be very small in macroscopic scale but very large in microscopic scale. Actually, there is an underlying fundamental symmetry between the quantum spacetime states and quantum matter states that we may call it quantum spacetime matter symmetry, or object-observer symmetry. In fact, an observer could be an object to another observer or vice versa and this actually is the essence of the principle of relativity.

Finally, it is expected that, given that the equivalence principle of quantum gravity and the energy scale dependent quantum spacetime concept are introduced under the double conformal symmetry of the quantum space-time-matter geometry, it is reasonable that other mathematical formalism can be used as well to arrive at the same conclusions. The mathematical treatment is not necessary restricted to the approach introduced in this book chapter.

Author details

Ho-Ming Mok
RHU, Hong Kong SAR, China

*Address all correspondence to: a8617104@graduate.hku.hk

References

[1] Mok HM. A possible solution to the cosmological constant problem by discrete space-time hypothesis. In: On the Nature of Dark Energy-Proceedings of the 18th Institut d'Astropysique de Paris Colloquium; Frontier Group. 2002. arXiv:astro-ph/0105513

[2] Mok HM. The solution of cosmological constant problem by discrete space-time at electroweak scale. In: Proceedings of the 12th Marcel Grossmann Meeting on General Relativity; 2009; Paris. Singapore: World Scientific; 2012

[3] Perlmutter S et al. Measurements of Ω and Λ from 42 high-redshift supernovae. The Astrophysical Journal. 1999;**517**:565

[4] Reiss AG et al. Type Ia supernova discoveries at z>1 from the Hubble Space Telescope: evidence for past deceleration and constraints on dark energy evolution. The Astrophysical Journal. 2004;**607**:665

[5] Mok HM. The relationship between the problems of cosmological constant and CP violation. In: Proceedings of the Eleventh International Symposium Frontier of Fundamental Physics; 2010; Paris. New York: American Institute of Physics; 2012

[6] Mok HM. Cosmological constant problem and equivalence principle of quantum gravity. In: Proceedings of the 13th Marcel Grossmann Meeting on General Relativity; 2012; Stockholm. Singapore: World Scientific; 2015

[7] Colella R et al. Observation of gravitationally induced quantum interference. Physical Review Letters. 1975;**34**:1472

[8] Adler R et al. Introduction to General Relativity. Tokyo: McGraw-Hill Kogakusha; 1975

[9] Rovelli C. Quantum Gravity. Cambridge: Cambridge University Press; 2004

[10] Rovelli C, Vidotto F. Covariant Loop Quantum Gravity An Elementary Introduction to Quantum Gravity and Spinfoam Theory. Cambridge: Cambridge University Press; 2015

[11] Tanabashi et al (Particle Data Group). Review of particle physics. 2018;**98**:030001

[12] Hooper D. The empirical case for 10-GeV dark matter. Dark Universe. 2012;**1**:1-23